Aristotle
The Physics.
Writings on Natural
Philosophy

science foundations the building blocks of life

This is a **FLAME TREE** book

FLAME TREE PUBLISHING
6 Melbray Mews, Fulham,
London SW6 3NS, United Kingdom
www.flametreepublishing.com

First published and copyright
© 2023 Flame Tree Publishing Ltd

This edition is based on an abridged version of the original
publication of *Physica*, by Aristotle, translated by R.P. Hardie
and R.K. Gaye and published by Oxford University Press in 1930
as Volume II of *The Works of Aristotle*.

23 25 27 26 24
1 3 5 7 9 10 8 6 4 2

ISBN: 978-1-80417-566-8

Cover and pattern art was created by Flame Tree Studio.

A copy of the CIP data for this book is available
from the British Library.

Designed and created in the UK | Printed and bound in China

Aristotle
The Physics.
Writings on Natural Philosophy

A New Introduction by Dr James Lees

Series Foreword by Professor Marika Taylor

FLAME TREE
CONCISE
EDITIONS

Contents

Series Foreword

S cience has always been a source of fascination, pushing the frontiers of knowledge. This series of books spans some of the most important scientific developments of all time. From the origins of life to the nature of the universe, these books showcase remarkable achievements of scientists through the ages and the limitless potential of human curiosity.

In this series we can follow the evolution of science from ancient times through to the modern era. Aristotle's writings cover a wide range of topics related to the natural world, including physics, metaphysics and biology. His work is notable for its depth and breadth as well as for its immense influence on the development of scientific thought in the Western world.

Copernicus and Newton were also great polymaths. Copernicus is best known for his work on the heliocentric nature of the solar system as presented in *On the Revolutions of the*

Heavenly Spheres but he also made substantial contributions to mathematics and medicine. Newton's *Principia* is recognized as one of the most important scientific works ever written. His theories of motion and gravity laid the foundations for modern physics and revolutionized our understanding of the 'Nature' of the world. Within mathematics, Newton developed calculus, as well as algebra, trigonometry and geometry.

Newton's *Principia* was originally read primarily by intellectuals. By contrast, Darwin's work *On the Origin of Species* immediately caused a significant stir in both scientific and public circles, particularly through its challenge to traditional beliefs. Darwin's theory of evolution was debated and refined, gradually becoming a foundational concept of modern biology.

At the beginning of the twentieth century, physics experienced two major paradigm shifts: relativity and quantum theory. Einstein's *Special and General Relativity* fundamentally changed our understanding of space, time and gravity. The works of Bohr and Planck explain the early development of quantum theory, starting from Planck's revolutionary concept of quanta.

Readers of this series can step back in time to explore how the foundations of science are described by its pioneers. The transformative nature of these works and their profound impacts on society can only inspire a sense of awe and wonder.

Professor Marika Taylor
University of Southampton

A New Introduction to Aristotle

The World's Most Famous Teacher

In reading a work of Aristotle (384–22 BCE) you are joining a class of many of the most eminent and influential figures throughout the ages. His body of work is perhaps the most well-known and best-read of any in history. To date, over 40 works survive (though the true authorship of some of these is disputed) in a compilation of knowledge that touches nearly every facet of existence.

Be it his works on ethics and politics or his foundational works on logic, the texts of Aristotle proved to be required reading for any person considering themself learned right through to the twentieth century. They were often held up as pinnacles of knowledge and used with an almost religious reverence. Here, however, we concern ourselves with some of his more worldly texts. In *Physica* (Physics, included in this volume) and the related *De Caelo et Mundo* (On the Heavens) Aristotle set out what would become some of the foundational ideas for science, shaping its progress for the next two millennia.

It is worth noting that even with its translation into something resembling modern parlance that the works of Aristotle do not make for light reading. The time and place into which Aristotle was born was a vastly different world to our own, and the tools and ideas that he had available to him were incredibly limited. It is for this reason that he has to work everything out from first principles, taking nothing for granted, and these ideas can be difficult for us to grapple with.

However, you do not receive titles like 'The Master of Those Who Know' from such luminaries as Dante Alighieri (1265–1321) without producing something truly special. In spending the time to read and digest his works, you can gain an understanding of a mind that sets the stage for all scientific inquiry to come, and touch a timeline of fellow students that includes kings, priests, scholars and change-makers throughout the world.

Science in 400 BCE

To be able to understand his work, it is important to envisage the world into which Aristotle was born. By Aristotle's birth, the walls of Troy had fallen some 800 years earlier and the

11

Mycenaean Greeks who toppled them had already passed into myth and legend. The second Peloponnesian war had led to Spartan dominance a couple of decades ago, but their grip was already beginning to slip.

Despite the city-state of Athens having lost its near total hegemony, it remained the centre of the Greek world for much of its politics, theatre and philosophy. Socrates (470–399 BCE) had been executed for 'impiety and corrupting the youth' and his star pupil Plato (*c.* 428–*c.* 348 BCE) had just set up 'The Academy' in 387 BCE as a centre of learning and scholarly conduct.

The science of the day looks almost indistinguishable to us now. Focused more on rational debates and deep consideration, those schooled in Socratic methods sought to think their way to all of life's answers. But here there is an important line that separates the Greek thinkers from those who came before. They were far from the first to ask questions, or even the first to try to answer with what we might now consider science (astronomical records have been unearthed from Babylon dating back to *c.* 2300 BCE), but they had begun to try to form structures of explanation for the world around them. This was called 'natural philosophy' and was a form of self-conscious inquiry into nature. It was a school of thought that looked at the things they saw on a daily basis and attempted to explain why they were like that. The Greeks of this time held complex views on religion, but this natural philosophy sought to examine those things of nature which are not of the gods.

Despite the cultural image of this time (largely informed by Renaissance paintings) of argumentative long-bearded men, many of the ideas that would define science had already begun to take shape. The Socratic method of deductive reasoning was a dominant form of logical argument, Pythagoras (*c.* 570–*c.* 495 BCE) had brought mathematics as a practical form of enquiry, and even early experiments had occurred such as Democritus (*c.* 460–*c.* 370 BCE) placing an upturned jar in a bath of water to show that air must contain something rather than nothing.

Aristotle the Man

It was into this world that in 384 BCE Aristotle was born. Almost nothing is known of Aristotle's early life. Born in the northern coastal city of Stagira, he was the son of a well-connected doctor who worked as the personal physician of the king of Macedon, Amyntas III (420–370 BCE). Likely growing up in and around the Macedonian court, his father tutored him in anatomy and medicine. However, both of his parents died while Aristotle was still a young boy, and he was left in the care of a friend who at the age of around 17 sent him off to join Plato's Academy.

It was here that Aristotle would become a true scholar, absorbing all the knowledge classical Greece had to offer and engaging in lively debates with his fellow pupils. He stayed there for nearly 20 years until Plato died. It is not clear exactly why he left, with assorted theories from his displeasure with the new leader of the Academy to simply deciding that he should see more of the world. In any case, he spent time in Anatolia and Lesbos furthering his knowledge with his friends and pupils.

His life took a turn in 343 BCE when Aristotle was invited by the then king of Macedon, Philip II (382–336 BCE), youngest son of Amyntas III (and fourth to succeed the throne from him), to tutor his young son Alexander III (356–323 BCE). The boy would go on to be known as 'Alexander the Great' for his grand conquests of the Achaemenid Persian Empire, taking control of a huge swathe of land from Greece and Egypt to the Indus River in modern Pakistan in just 13 years.

Throughout the future leader's early life and some of his early reign, Aristotle worked as an instructor in logic and natural philosophy both as a personal tutor but also as the head of the Macedon's royal academy. Even after leaving to return to Athens in 335 BCE, he remained a confidant and advisor to Alexander, until relations strained some time before the king's death.

Once back in his scholastic stomping ground, he established his own academic institution, 'The Lyceum', to rival that of his old tutor. It was here that Aristotle delivered his now well-considered and honed thoughts that became his famous works. Known for his tendency to continually walk while delivering his lectures, the school of thought became known as the peripatetic school (meaning 'given to walking about'). His lectures, which varied in length and topic, were dutifully recorded by his students, who placed the scrolls in the library where they were eventually copied and spread around the world.

Sadly, this situation did not last, and when Alexander the Great died in 323 BCE, causing a collapse of the empire into a kind of civil war, Aristotle became a focal point of anti-Macedonian sentiment. This forced him to flee from Athens to the islands of Euboea where he died shortly after in 322 BCE.

Aristotle's Body of Work

What, then, was so ground-breaking about Aristotle's work? Why was it his scholarship that became the bedrock of almost everything to come afterwards? There are, of course, some practical considerations to his popularity. Coming directly from the famous lineage of the preeminent thinkers Socrates and Plato placed him as their obvious successor, and being widely known as the teacher of the man who had conquered the then largest ever empire must have been helpful. However, the true secret of his enduring success came not from his connections, but his ability to bring together the work of those before him into one coherent world view, which he then rigorously analysed.

To discuss any one part of Aristotle's work is to touch upon many of his others. While this book contains only *Physics*, which formulates the previously mentioned complete world view that is the basis of his later work, considering his other work lets us understand how and why he was able to produce them.

15

One of his greatest achievements was the creation and study of 'formal logic', which turned rational debates into a kind of maths problem. Taking a famous example, if we accept the following two statements:

All Men are Mortal
Socrates is a Man

We must therefore conclude that:

Socrates is Mortal

While such a problem might seem trivial to us today, at the time this formalisation was positively innovative. It allows for statements to be made on the mortality of Socrates knowing nothing more than he is a man, and while this example is basic, the reasoning behind it seems so remarkably simple precisely because it is something we use every day. Other (now seemingly trifling) concepts that Aristotle formalised included the law of non-contradiction, stating that a given preposition must be either true or false.

This conception of logic allowed him to form something of a proto-empirical world view. Importantly, this led him to two major ideas that were the basis of his work. The first was a strict separation of the natural and the mythical. Like many of his peers, Aristotle saw the gods in all things, but in his examinations of the world broadly left them out. While this in itself was not particularly

16

unusual, he went further than most by saying that things like the wind or other unseen powers, such as volcanoes, were not directly caused by the gods but rather natural phenomena. Secondly, was his belief in first observing the world using our senses and then explaining why the things he saw were happening. Doing it in any other way might seem ludicrous to us now, but at the time many thought that the entire universe could be explained with enough consideration and nothing else.

Later works of Aristotle included examinations on geology and biology. While these included plenty of postulating, much of them is dedicated to a simple collection of knowledge. Almost as a kind of encyclopaedic exercise, he sought to record the various kinds of plants, animals and land formations. Then he attempted to categorize them using his four qualities: hot, cold, wet and dry. He also categorized things by their 'souls'; the vegetative soul that led to reproduction and growth, and existed in all living things; the sensitive soul that allowed for movement and sensing, which existed in all animals; and finally, the rational soul that allowed for rational thinking and was present only in humans. This led to him creating a hierarchy of the simplest to the most complex, placing the less soulful, cold and dry plants at the bottom, and the soul – rich, hot, wet man – at the top. However, his idea that human females are produced by cold fluids and males by hot ones belies how his ideas on politics informed his ideas of natural philosophy.

Aristotle wrote not only on biology, geology, logic and natural philosophy; he also wrote on botany, economics, politics, rhetoric, poetry, ethics and even dreams. It was a time when an educated person could touch on almost every topic in an important way and, in doing so, they were all interlinked. So even as we examine just a couple of his texts, we must consider them as part of a greater body of work, which offers a view into one of the greatest minds of the era.

Sadly, many of the books written by and from the teachings of Aristotle do not survive to the modern age. Time can be cruel to texts, even those as well regarded and oft-copied as Aristotle, and by the most extreme estimates, up to three-quarters of what he produced has been lost. There exist references in the works of later scholars and even some fragments (although of suspicious authenticity) that tell us of what once was. However, until a serendipitous discovery on some ancient dig site or long-forgotten corner of an aged library, we must work with what we have.

The Four Causes

Physics in many ways can be considered the first of his natural philosophy books. That is because it sets up how Aristotle believed the heavens and the earth function, which serves as a foundation for all of his other ideas. *Physics* focuses on the principles of natural things. Why are things the way they are? Most importantly it looks at why things change and move, rather than remaining inert. It spends much of its time defining the world around us and giving us conceptions of space and time for things to exist within.

Of most interest within the work is Aristotle's 'Four Causes'. For this, he defines everything as being either of potentiality or actuality: that everything has a natural end state that it is striving to reach, which he called 'telos', and all things have a potentiality until they reach this end state, at which point they have actuality. So, the seed of a tree has potentiality, and once fully grown into its final state of telos, it has gained actuality. In understanding the question of 'why' any given object reaches its final state, Aristotle defines the four causes.

Material Cause

This is the material that the thing is made of that will define its properties. For example, being made of stone would make something heavy or being made of wood might allow it to float.

Formal Cause

The formal cause is the definition or essence of a thing or its shape. A wheel will turn because it is round, and a table will stay upright because of its legs.

Efficient Cause

This is perhaps closer to what we would today define as a cause. This is a definition of the actual agent of the change. So, the effective cause of a cart would be the cartwright who built it and, by extension, the efficient cause of the cartwright (or any person) would be their parents. This does not have to be a person. For example, a river would be the efficient cause of a leaf floating downstream.

Final Cause

The final cause is the answer to why something exists, its purpose. So, for a car this would be to be driven, or for an apple tree this is to produce edible fruits.

Aristotle did not see all of these causes as totally separate, but rather that a mix of them was required to explain why something was the way it was. For example, a lyre might be brown and solid because it was made of wood (material cause), it was shaped the way it was because that was the shape that allowed it to string the instrument (formal cause), it was in said shape because it had been crafted as such by a craftsman (efficient cause), and then it produced music (final cause).

Aristotle places a primacy on the final cause, arguing that to reach that end some of the other causes already need to be a certain way. For example, at the very end of book two, he states:

> '*For if one defines the operation of sawing as being a certain kind of dividing, then this cannot come about unless the saw has teeth of a certain kind; and these cannot be unless it is of iron. For in the definition too there are some parts that are, as it were, its matter.*'

This way of looking at the world may now seem simplistic and deterministic at best. If we consider a seed becoming a tree, we can work out its causes, but they would not be able to account for if the tree grew tall or short. The very notion of a telos as some predetermined end state also strikes of a kind of godly idealism. How could we know if or when a tree has reached its telos as it will continue to grow for its entire life, even if very slowly? However, at the time, it was one of the first fully universal ways of describing the reasons behind anything and everything. That said, even by his own admission, it can be difficult to fully work out what the causes for everything are and he left plenty of room for others to continue his work.

The Aristotelian World

As mentioned previously, it can be difficult when you don't consider Aristotle as a whole. As such if we skip forward in his bibliography to *On the Heavens*, here Aristotle gives us his conception of the natural world, the Greek word *physis* literally meaning 'nature'. He takes the four elements (earth, air, wind and fire) common in Greek natural philosophy and describes a world order as a hierarchy of spheres. In a perfect world this would exist as a sphere of earth surrounded by water, which itself was surrounded by a sphere of air, which sat in a sphere of fire, all of which was within an unchanging sphere of aether.

21

The desire, then, for these elements to return to their natural position is the driving force behind most of the change seen in the world and helps to describe in some way the essence of the four causes. It explained why lakes and rivers sat on top of the ground and why flames leap up into the sky. Everything else is made of a mixture of these elements. Heavier things contain more earth, so they fall to the ground faster. Trees contain a significant amount of both earth and air, which causes them to simultaneously reach into the sky above and down into the ground below.

On the Heavens also concerns the outer sphere of an unchanging substance called 'aether'. It gives a model of a geocentric universe (one with the earth at its centre), where the heavens are fixed and capable only of rotating around us. In many ways, it is a significant break with those who came before, especially Plato and his ideas of a perfect heaven. For this reason, much of the book is dedicated to refuting previous ideas and supplanting them with his own.

22

Aristotle's Legacy

It seems clear, then, that Aristotle's natural philosophy bears little resemblance to the science that we perform today. Nearly every facet of his work is completely wrong, indeed even by the early 500s some scholars were already writing scathing critiques of his science. His shortcomings are evident in such famed statements as that in *History of Animals*, which said:

> *'Males have more teeth than females: in humans, in sheep, in goats, and in pigs; and in the other species which have yet to be observed.'*

This is demonstrably untrue and would take only the slightest amount of examination to prove. So, the question is why this now potentially juvenile way of doing science and questioning our world was dominant for nearly 2,000 years and is even today held in high esteem by philosophers and scientists alike.

After Aristotle's death the Lyceum carried on under his students, but as the generations changed, its focus shifted. No longer did the scholars within its halls look to discover new knowledge but rather to learn and understand Aristotle's work. During the Roman era Athens was no longer the academic centre of the world and Aristotle's work became just another part of the empire's lively scholarly canon, if a well-regarded one. Through the rubble and flames of the collapse of the Roman Empire, Aristotle survived thanks to the scholars of the Byzantines, who created numerous copies and made efforts to spread them around.

In the middle of the first century, the Latin west began to lose interest in the works of the Greeks, and Aristotle disappeared almost entirely from shelves and classrooms. Fortunately, the Muslim scholar Al-Kindi (*c.* 801–73) translated many of his works and introduced them to the Eastern world, where he became the most respected Western thinker throughout the entirety of the Islamic Golden Age. This led to his ideas influencing much of Arabic philosophy and science, gaining him the revered title of 'First Teacher'.

It was not until the revival and 'rediscovery' of the classical world (largely re-translated from Arabic) that marked the early Renaissance that Aristotle regained his importance in the Latin West. Now once again, his ideas of both philosophy and science were learned by every educated person from France to Lithuania and became increasingly widely available thanks to the invention and use of the printing press. So, for a few hundred years more Aristotle and his peers reassumed the position of nearly unquestioned sources of knowledge that they had enjoyed in antiquity.

It was not long, however, before the cracks really began to show. Through the work of luminaries such as Nicholas Copernicus (1473–1543), Andreas Vesalius (1514–64) and Galileo Galilei (1564–1642), the idea of performing experiments and first-hand observations to gain new knowledge began to overtake simply referring to the works of the old masters as gospel. Despite this wane in his status,

Aristotle and his contemporaries remained vitally important texts. Their reading remained practically compulsory for any self-respecting scientist right through until the early twentieth century.

The power of Aristotle's work does not come from a single place. To be able to survive and thrive across the years, and through a multitude of wholly distinct cultures, it has had to be so many different things to so many different people. On the practical side, his fame within his time ensured that enough copies of his work were created so that they could be widely circulated and read, allowing them to continue to survive where the works of so many others were lost completely.

However, it would be drastically unfair to reduce his longevity down to some simple materialistic reasons. Within his works he also brought together the ideas of many of those before him, indeed this was to such an extent that many, both at the time and later, declared that there was simply no need to learn the lessons of the earlier Greeks, as he covered it all.

We must also not underestimate the power of being the progenitor of so many things. His body of work touched on so many subjects, so early in their development, that even today it is difficult not to see his fingerprints in the form that they take. This foundational role meant that understanding later work was often predicated on already being familiar with the works of Aristotle.

Finally, it is the holistic view that Aristotle takes in trying to understand all nature that gives him his importance.

Despite the now obvious flaws in his reasoning, he looked at the world and thought to break it down into more understandable components and hierarchies. He was one of the first to say that not only was it possible to question everything, but that with enough effort it would be possible to find the answer, too. And it is this, despite everything that has happened in the millennia since his death, that remains the fundamental central pillar of science.

A Note on this Edition

This concise edition presents a somewhat more digestible version of Aristotle's *Physics*, unfettered by the copious notes, mostly added by later scholars and translators, which are highly academic and lay heavy reference on other texts as a wider reading source. We also omit the final book (Book VIII) as it is significantly more religious and philosophical than many of the other parts of the text and its removal does not detract from the broader scientific understanding.

Dr James Lees

Aristotle
The Physics

Translated by R.P. Hardie
and R.K. Gaye

Book I

The Scope and Method of This Book

When the objects of an inquiry, in any department, have principles, conditions, or elements, it is through acquaintance with these that knowledge, that is to say scientific knowledge, is attained. For we do not think that we know a thing until we are acquainted with its primary conditions or first principles, and have carried our analysis as far as its simplest elements. Plainly therefore in the science of Nature, as in other branches of study, our first task will be to try to determine what relates to its principles.

The natural way of doing this is to start from the things which are more knowable and obvious to us and proceed towards those which are clearer and more knowable by Nature; for the same things are not 'knowable relatively to us' and 'knowable' without qualification. So in the present inquiry we must follow this method and advance from what is more obscure by nature, but clearer to us, towards what is more clear and more knowable by nature.

Now what is to us plain and obvious at first is rather confused masses, the elements and principles of which become known to us later by analysis. Thus we must advance from generalities to particulars; for it is a whole that is best known to sense-perception, and a generality is a kind of whole, comprehending many things within it, like parts. Much the same thing happens in the relation of the name to the formula. A name, e.g. 'round', means vaguely a sort of whole: its definition analyses this into its particular senses. Similarly a child begins by calling all men 'father', and all women 'mother', but later on distinguishes each of them.

The Problem: The Number and Character of the First Principles of Nature

The principles in question must be either (*a*) one or (*b*) more than one. If (*a*) one, it must be either (i) motionless, as Parmenides and Melissus assert, or (ii) in motion, as the physicists hold, some declaring air to be the first principle, others water.

If (*b*) more than one, then either (i) a finite or (ii) an infinite plurality. If (i) finite (but more than one), then either two or three or four or some other number. If (ii) infinite, then either as Democritus believed one in kind, but differing in shape or form; or different in kind and even contrary.

A similar inquiry is made by those who inquire into the number of *existents*: for they inquire whether the ultimate constituents of existing things are one or many, and if many, whether a finite or an infinite plurality. So they too are inquiring whether the principle or element is one or many.

Now to investigate whether Being is one and motionless is not a contribution to the science of Nature. For just as the geometer has nothing more to say to one who denies the principles of his science – this being a question for a different science or for one common to all – so a man investigating *principles* cannot argue with one who denies their existence. For if Being is just one, and one in the way mentioned, there is a principle no longer, since a principle must be the principle of some thing or things.

To inquire therefore whether Being is one in this sense would be like arguing against any other position maintained for the sake of argument (such as the Heraclitean thesis, or such a thesis as that Being is one man) or like refuting a merely contentious argument – a description which applies to the arguments both of Melissus and of Parmenides: their premisses are false and their conclusions do not follow. Or rather the argument of Melissus is gross and palpable and offers no difficulty at all: accept one ridiculous proposition and the rest follows – a simple enough proceeding.

We physicists, on the other hand, must take for granted that the things that exist by nature are, either all or some of them, in motion – which is indeed made plain by induction. Moreover, no man of science is bound to solve every kind of

30

difficulty that may be raised, but only as many as are drawn falsely from the principles of the science: it is not our business to refute those that do not arise in this way: just as it is the duty of the geometer to refute the squaring of the circle by means of segments, but it is not his duty to refute Antiphon's proof. At the same time the holders of the theory of which we are speaking do incidentally raise physical questions, though Nature is not their subject: so it will perhaps be as well to spend a few words on them, especially as the inquiry is not without scientific interest.

Reality Is Not One in the Way that Parmenides and Melissus Supposed

The most pertinent question with which to begin will be this: In what sense is it asserted that all things are one? For 'is' is used in many senses. Do they mean that all things 'are' *substance* or *quantities* or *qualities*? And, further, are all things one substance – one man, one horse, or one soul – or quality and that one and the same – white or hot or something of the kind? These are all very different doctrines and all impossible to maintain.

For if *both* substance and quantity-and-quality are, then, whether these exist independently of each other or not, Being will be many.

If on the other hand it is asserted that all things are quality or quantity, then, whether substance exists or not, an absurdity results, if indeed the impossible can properly be called absurd. For none of the others can exist independently: substance alone is independently: for everything is predicated of substance as subject. Now Melissus says that Being is infinite. It is then a quantity. For the infinite is in the category of quantity, whereas substance or quality or affection cannot be infinite except through a concomitant attribute, that is, if at the same time they are also quantities. For to define the infinite you must use quantity in your formula, but not substance or quality. If then Being is both substance and quantity, it is two, not one: if only substance, it is not infinite and has no magnitude; for to have that it will have to be a quantity.

Again, 'one' itself, no less than 'being', is used in many senses, so we must consider in what sense the word is used when it is said that the All is one.

Now we say that (*a*) the continuous is one or that (*b*) the indivisible is one, or (*c*) things are said to be 'one', when their essence is one and the same, as 'liquor' and 'drink'.

If (*a*) their One is one in the sense of continuous, it is many, for the continuous is divisible *ad infinitum*.

There is, indeed, a difficulty about part and whole, perhaps not relevant to the present argument, yet deserving consideration on its own account – namely, whether the part and the whole are one or more than one, and how they can be one or many, and, if they are more than one, in what sense they are more than one. (Similarly with the parts of wholes which are not continuous.) Further, if each of the two parts is indivisibly one with the whole, the difficulty arises that they will be indivisibly one with each other also.

But to proceed: If (*b*) their One is one as indivisible, nothing will have quantity or quality, and so the one will not be infinite, as Melissus says – nor, indeed, limited, as Parmenides says, for though the limit is indivisible, the limited is not.

But if (*c*) all things are one in the sense of having the same definition, like 'raiment' and 'dress', then it turns out that they are maintaining the Heraclitean doctrine, for it will be the same thing 'to be good' and 'to be bad', and 'to be good' and 'to be not good', and so the same thing will be 'good' and 'not good', and man and horse; in fact, their view will be, not that all things are one, but that they are nothing; and that 'to be of such-and-such a quality' is the same as 'to be of such-and-such a size'.

Even the more recent of the ancient thinkers were in a pother lest the same thing should turn out in their hands both one and many. So some, like Lycophron, were led to omit 'is', others to change the mode of expression and say 'the man has been whitened' instead of 'is white', and 'walks' instead of 'is walking', for fear that if they added the word 'is' they should be making the one to be many – as if 'one' and 'being' were always used in one and the same sense. What 'is' may be many either in definition (for example 'to be white' is one thing, 'to be musical' another, yet the same thing may be both, so the one is many) or by division, as the whole and its parts.

On this point, indeed, they were already getting into difficulties and admitted that the one was many – as if there was any difficulty about the same thing being both one and many, provided that these are not opposites; for 'one' may mean either 'potentially one' or 'actually one'.

Refutation of Their Arguments

If, then, we approach the thesis in this way it seems impossible for all things to be one. Further, the arguments they use to prove their position are not difficult to expose. For both of them reason contentiously – I mean both Melissus and Parmenides. [Their premisses are false and their conclusions do not follow. Or rather the argument of Melissus is gross and palpable and offers no difficulty at all: admit one ridiculous proposition and the rest follows – a simple enough proceeding.]

The fallacy of Melissus is obvious. For he supposes that the assumption 'what has come into being always has a beginning' justifies the assumption 'what has not come into being has no beginning'. Then this also is absurd, that in every case there should be a beginning of the *thing* – not of the time and not only in the case of coming to be in the full sense but also in the case of coming to have a quality – as if change never took place suddenly. Again, does it follow that Being, if one, is motionless? Why should it not move, the whole of it within itself, as parts of it do which arc unities, e.g. this water? Again, why is qualitative change impossible? But, further, Being cannot be one in form, though it may be in what it is made of. (Even 20 some of the physicists hold it to be one in the latter way, though not in the former.) Man obviously differs from horse in form, and contraries from each other.

34

The same kind of argument holds good against Parmenides also, besides any that may apply specially to his view: the answer to him being that '*this* is not true' and '*that* does not follow'. His assumption that one is used in a single sense only is false, because it is used in several. His conclusion does not follow, because if we take only white things, and if 'white' has a single meaning, none the less what is white will be many and not one. For what is white will not be one either in the sense that it is continuous or in the sense that it must be defined in only one way. 'Whiteness' will be different from 'what has whiteness'. Nor does this mean that there is anything that can exist separately, over and above what is white. For 'whiteness' and 'that which is white' differ in definition, not in the sense that they are things which can exist apart from each other. But Parmenides had not come in sight of this distinction.

It is necessary for him, then, to assume not only that 'being' has the same meaning, of whatever it is predicated, but further that it means (1) what *just is* and (2) what is *just one*.

It must be so, for (1) an attribute is predicated of some subject, so that the subject to which 'being' is attributed will not be, as it is something different from 'being'. Something, therefore, which is not will be. Hence 'substance' will not be a predicate of anything else. For the subject cannot be a *being*, unless 'being' means several things, in such a way that each *is* something. But *ex hypothesi* 'being' means only one thing.

If, then, 'substance' is not attributed to anything, but other things are attributed to it, how does 'substance' mean what is rather than what is not? For suppose that 'substance' is also 'white'. Since the definition of the latter is different (for being cannot even be attributed to white, as nothing is which is not 'substance'), it follows that 'white' is not-being – and that not in the sense of a particular not-being, but in the sense that it is not at all. Hence 'substance' is not; for it is true to say that it is white, which we found to mean not-being. If to avoid this we say that even 'white' means substance, it follows that 'being' has more than one meaning.

In particular, then, Being will not have magnitude, if it is substance. For each of the two parts must *be* in a different sense.

(2) Substance is plainly divisible into other substances, if we consider the mere nature of a definition. For instance, if 'man' is a substance, 'animal' and 'biped' must also be substances. For if not substances, they must be attributes – and if attributes, attributes either of (*a*) man or of (*b*) some other subject. But neither is possible.

(*a*) An attribute is either that which may or may not belong to the subject or that in whose definition the subject of which it is an attribute is involved. Thus 'sitting' is an example of a separable attribute, while 'snubness' contains the definition of 'nose', to which we attribute snubness. Further, the definition of the whole is not contained in the definitions of the contents or elements of the definitory formula; that of 'man' for instance in 'biped', or that of 'white man' in 'white'. If then this is so, and if 'biped' is supposed to be an attribute of 'man', it must be either separable, so that 'man' might possibly not be 'biped', or the definition of 'man' must come into the definition of 'biped' – which is impossible, as the converse is the case.

(*b*) If, on the other hand, we suppose that 'biped' and 'animal' are attributes not of man but of something else, and are not each of them a substance, then 'man' too will be an attribute of something else. But we must assume that substance is *not* the attribute of anything, and that the subject of which both 'biped' and 'animal' and each separately are predicated is the subject also of the complex 'biped animal'.

Are we then to say that the All is composed of indivisible substances? Some thinkers did, in point of fact, give way to both arguments. To the argument that all things are one if being means one thing, they conceded that not-being is; to that from bisection, they yielded by positing atomic magnitudes. But obviously it is not true that if being means one thing, and cannot at the same time mean the contradictory of this, there will be nothing which is not, for even if what is not cannot *be* without qualification, there is no reason why it should not be a particular not-being. To say that all things will be one, if there is nothing besides Being itself, is absurd. For who understands 'being itself' to be anything but a particular substance? But if this is so, there is nothing to prevent there being many beings, as has been said.

It is, then, clearly impossible for Being to be one in this sense.

Statement and Examination of the Opinions of the Natural Philosophers

The physicists on the other hand have two modes of explanation.

The first set make the underlying body one – either one of the three or something else which is denser than fire and rarer

37

than air – then generate everything else from this, and obtain multiplicity by condensation and rarefaction. Now these are contraries, which may be generalized into 'excess and defect'. (Compare Plato's 'Great and Small' – except that he makes these his matter, the one his form, while the others treat the one which underlies as matter and the contraries as differentiae, i.e. forms).

The second set assert that the contrarieties are contained in the one and emerge from it by segregation, for example Anaximander and also all those who assert that 'what is' is one and many, like Empedocles and Anaxagoras; for they too produce other things from their mixture by segregation. These differ, however, from each other in that the former imagines a cycle of such changes, the latter a single series. Anaxagoras again made both his 'homoeomerous' substances and his contraries infinite in multitude, whereas Empedocles posits only the so-called elements.

The theory of Anaxagoras that the principles are infinite in multitude was probably due to his acceptance of the common opinion of the physicists that nothing comes into being from not-being. For this is the reason why they use the phrase 'all things were together' and the coming into being of such and such a kind of thing is reduced to change of quality, while some spoke of combination and separation. Moreover, the fact that the contraries proceed from each other led them to the conclusion. The one, they reasoned, must have already existed in the other; for since everything that comes into being must arise either from what is or from what is not, and it is impossible for it to arise from what is not (on this point all the physicists agree), they thought that the truth of the alternative necessarily followed, namely that things come into being out of existent things, i.e. out of things already present, but imperceptible to our senses because of the smallness of their bulk. So they assert that everything has been mixed in everything, because they saw everything arising out of everything.

But things, as they say, appear different from one another and receive different names according to the nature of the particles which are numerically predominant among the innumerable constituents of the mixture. For nothing, they say, is purely and entirely white or black or sweet, bone or flesh, but the nature of a thing is held to be that of which it contains the most.

Now (1) the infinite *qua* infinite is unknowable, so that what is infinite in multitude or size is unknowable in quantity, and what is infinite in variety of kind is unknowable in quality. But the principles in question are infinite both in multitude and in kind. Therefore it is impossible to know things which are composed of them; for it is when we know the nature and quantity of its components that we suppose we know a complex.

Further (2) if the parts of a whole may be of any size in the direction either of greatness or of smallness (by 'parts' I mean components into which a whole can be divided and which are actually present in it), it is necessary that the whole thing itself may be of any size. Clearly, therefore, since it is impossible for an animal or plant to be indefinitely big or small, neither can its parts be such, or the whole will be the same. But flesh, bone, and the like are the parts of animals, and the fruits are the parts of plants. Hence it is obvious that neither flesh, bone, nor any such thing can be of indefinite size in the direction either of the greater or of the less.

Again (3) according to the theory all such things are already present in one another and do not come into being but are constituents which are separated out, and a thing receives its designation from its chief constituent. Further, anything may come out of anything – water by segregation from flesh and flesh from water. Hence, since every finite body is exhausted by the repeated abstraction of a finite body, it seems obviously to follow that everything *cannot* subsist in everything else. For let flesh be extracted from water and again more flesh be produced from the remainder by repeating the process of separation: then, even though the quantity separated out will continually decrease, still it will not fall below a certain magnitude. If, therefore, the process comes to an end, everything will not be in everything else (for there will be no flesh in the remaining water); if on the other hand it does not, and further extraction is always possible, there will be an infinite multitude of finite equal particles in a finite quantity – which is impossible. Another proof may be added: Since every body must diminish in size when something is taken from it, and flesh is quantitatively definite in respect both of greatness and smallness, it is clear that from the minimum quantity of flesh no body can be separated out; for the flesh left would be less than the minimum of flesh.

Lastly (4) in each of his infinite bodies there would be already present infinite flesh and blood and brain – having a distinct existence, however, from one another, and no less real than the infinite bodies, and each infinite: which is contrary to reason.

The statement that complete separation never will take place is correct enough, though Anaxagoras is not fully aware of what it means. For affections are indeed inseparable. If then colours and states had entered into the mixture, and if separation took place, there would be a 'white' or a 'healthy' which was nothing *but* white or healthy, i.e. was not the predicate of a subject. So his 'Mind' is an absurd person aiming at the impossible, if he is supposed to wish to separate them, and it is impossible to do so, both in respect of quantity and of quality – of quantity, because there is no minimum magnitude, and of quality, because affections are inseparable.

Nor is Anaxagoras right about the coming to be of homogeneous bodies. It is true there is a sense in which clay is divided into pieces of clay, but there is another in which it is not. Water and air are, and are generated, 'from' each other, but not in the way in which bricks come 'from' a house and again a house 'from' bricks; and it is better to assume a smaller and finite number of principles, as Empedocles does.

The Principles Are Contraries

All thinkers then agree in making the contraries principles, both those who describe the All as one and unmoved (for even Parmenides treats hot and cold as principles under the names

41

of fire and earth) and those too who use the rare and the dense. The same is true of Democritus also, with his plenum and void, both of which exist, he says, the one as being, the other as not-being. Again he speaks of differences in position, shape, and order, and these are genera of which the species are contraries, namely, of position, above and below, before and behind; of shape, angular and angle-less, straight and round.

It is plain then that they all in one way or another identify the contraries with the principles. And with good reason. For first principles must not be derived from one another nor from anything else, while everything has to be derived from them. But these conditions are fulfilled by the primary contraries, which are not derived from anything else because they are primary, nor from each other because they are contraries.

But we must see how this can be arrived at as a reasoned result, as well as in the way just indicated.

Our first presupposition must be that in nature nothing acts on, or is acted on by, any other thing at random, nor may anything come from anything else, unless we mean that it does so in virtue of a concomitant attribute. For how could 'white' come from 'musical', unless 'musical' happened to be an attribute of the not-white or of the black? No, 'white' comes from 'not-white' – and not from *any* 'not-white', but from black or some intermediate colour. Similarly, 'musical' comes to be from 'not-musical', but not from *any* thing other than musical, but from 'unmusical' or any intermediate state there may be.

Nor again do things pass into the first chance thing; 'white' does not pass into 'musical' (except, it may be, in

42

virtue of a concomitant attribute), but into 'notwhite' – and not into any chance thing which is not white, but into black or an intermediate colour; 'musical' passes into 'not-musical' – and not into any chance thing other than musical, but into 'unmusical' or any intermediate state there may be.

The same holds of other things also: even things which are not simple but complex follow the same principle, but the opposite state has not received a name, so we fail to notice the fact. What is in tune must come from what is not in tune, and *vice versa*; the tuned passes into untunedness – and not into any untunedness, but into the corresponding opposite. It does not matter whether we take attunement, order, or composition for our illustration; the principle is obviously the same in all, and in fact applies equally to the production of a house, a statue, or any other complex. A house comes from certain things in a certain state of separation instead of conjunction, a statue (or any other thing that has been shaped) from shapelessness – each of these objects being partly order and partly composition.

If then this is true, everything that comes to be or passes away comes from, or passes into, its contrary or an intermediate state. But the intermediates are derived from the contraries – colours, for instance, from black and white. Everything, therefore, that comes to be by a natural process is either a contrary or a product of contraries.

Up to this point we have practically had most of the other writers on the subject with us, as I have said already: for all of them identify their elements, and what they call their principles,

43

with the contraries, giving no reason indeed for the theory, but constrained as it were by the truth itself. They differ, however, from one another in that some assume contraries which are more primary, others contraries which are less so: some those more knowable in the order of explanation, others those more familiar to sense. For some make hot and cold, or again moist and dry, the conditions of becoming; while others make odd and even, or again Love and Strife; and these differ from each other in the way mentioned.

Hence their principles are in one sense the same, in another different; different certainly, as indeed most people think, but the same inasmuch as they are analogous; for all are taken from the same table of columns, some of the pairs being wider, others narrower in extent. In this way then their theories are both the same and different, some better, some worse; some, as I have said, take as their contraries what is more knowable in the order of explanation, others what is more familiar to sense. (The universal is more knowable in the order of explanation, the particular in the order of sense: for explanation has to do with the universal, sense with the particular.) 'The great and the small,' for example, belong to the former class, 'the dense and the rare' to the latter.

It is clear then that our principles must be contraries.

The Principles Are Two, or Three, in Number

The next question is whether the principles are two or three or more in number.

One they cannot be, for there cannot be one contrary. Nor can they be innumerable, because, if so, Being will not be knowable: and in any one genus there is only one contrariety, and substance is one genus: also a finite number is sufficient, and a finite number, such as the principles of Empedocles, is better than an infinite multitude; for Empedocles professes to obtain from his principles all that Anaxagoras obtains from his innumerable principles. Lastly, some contraries are more primary than others, and some arise from others – for example sweet and bitter, white and black – whereas the principles must always remain principles.

This will suffice to show that the principles are neither one nor innumerable.

Granted, then, that they are a limited number, it is plausible to suppose them more than two. For it is difficult to see how either density should be of such a nature as to act in any way on rarity or rarity on density. The same is true of any other pair of contraries; for Love does not gather Strife together and make things out of it, nor does Strife make anything out of Love, but both act on a third thing different from both. Some indeed assume more than one such thing from which they construct the world of nature.

Other objections to the view that it is not necessary to assume a third principle as a substratum may be added.

(1) We do not find that the contraries constitute the *substance* of any thing. But what is a first principle ought not to be the *predicate* of any subject. If it were, there would be a principle of the supposed principle: for the subject is a principle,

45

and prior presumably to what is predicated of it. Again (2) we hold that a substance is not contrary to another substance. How then can substance be derived from what are not substances? Or how can non-substance be prior to substance?

If then we accept both the former argument and this one, we must, to preserve both, assume a third somewhat as the substratum of the contraries, such as is spoken of by those who describe the All as one nature – water or fire or what is intermediate between them. What is intermediate seems preferable; for fire, earth, air, and water are already involved with pairs of contraries. There is, therefore, much to be said for those who make the underlying substance different from these four; of the rest, the next best choice is air, as presenting sensible differences in a less degree than the others; and after air, water. All, however, agree in this, that they differentiate their One by means of the contraries, such as density and rarity and more and less, which may of course be generalized, as has already been said, into excess and defect. Indeed this doctrine too (that the One and excess and defect are the principles of things) would appear to be of old standing, though in different forms; for the early thinkers made the two the active and the one the passive principle, whereas some of the more recent maintain the reverse.

To suppose then that the elements are three in number would seem, from these and similar considerations, a plausible view, as I said before. On the other hand, the view that they are more than three in number would seem to be untenable.

For the one substratum is sufficient to be acted on; but if we have four contraries, there will be two contrarieties, and we shall have to suppose an intermediate nature for each pair separately. If, on the other hand, the contrarieties, being two, can generate from each other, the second contrariety will be superfluous. Moreover, it is impossible that there should be more than one *primary* contrariety. For substance is a single genus of being, so that the principles can differ only as prior and posterior, *not* in genus; in a single genus there is always a single contrariety, all the other contrarieties in it being held to be reducible to one.

It is clear then that the number of elements is neither one nor more than two or three; but whether two or three is, as I said, a question of considerable difficulty.

The Number and Nature of the Principles

We will now give our own account, approaching the question first with reference to becoming in its widest sense: for we shall be following the natural order of inquiry if we speak first of common characteristics, and then investigate the characteristics of special cases.

We say that one thing comes to be from another thing, and one sort of thing from another sort of thing, both in the case of simple and of complex things. I mean the following. We can say (1) the 'man becomes musical', (2) what is 'not-musical becomes musical', or (3) the 'not-musical man becomes a musical man'. Now what becomes in (1) and (2) – 'man' and 'not musical' – I call *simple*, and what each becomes – 'musical' – simple also. But when we say the 'not-musical man becomes a musical man', both what becomes and what it becomes are *complex*.

As regards one of these simple 'things that become' we say not only 'this becomes so-and-so', but also 'from being this, comes to be so-and-so', as 'from being not-musical comes to be musical'; as regards the other we do not say this in all cases, as we do not say (1) 'from being a man he came to be musical' but only 'the man became musical'.

When a 'simple' thing is said to become something, in one case (1) it survives through the process, in the other (2) it does not. For the man remains a man and is such even when he becomes musical, whereas what is not musical or is unmusical does not continue to exist, either simply or combined with the subject.

These distinctions drawn, one can gather from surveying the various cases of becoming in the way we are describing that, as we say, there must always be an underlying something. namely that which becomes, and that this, though always one numerically, in form at least is not one. (By that I mean that it can be described in different ways.) For 'to be man' is not the same as 'to be unmusical', One part survives, the other does not: what is not an opposite survives (for 'man' survives), but 'not-musical' or 'unmusical' does not survive, nor does the compound of the two, namely 'unmusical man'.

We speak of 'becoming that from this' instead of 'this becoming that' more in the case of what does not survive the change – 'becoming musical from unmusical', not 'from man' – but there are exceptions, as we sometimes use the latter form of expression even of what survives; we speak of 'a statue coming to be from bronze', not of the 'bronze becoming a statue'. The change, however, from an opposite which does not survive is described indifferently in both ways, 'becoming that from this' or 'this becoming that'.

We say both that 'the unmusical becomes musical', and that 'from unmusical he becomes musical'. And so both forms are used of the complex, 'becoming a musical man from an unmusical man', and 'an unmusical man becoming a musical man'.

But there are different senses of 'coming to be'. In some cases we do not use the expression 'come to be', but 'come to be so-and-so'. Only substances are said to 'come to be' in the unqualified sense.

Now in all cases other than substance it is plain that there must be some subject, namely, that which becomes. For we know that when a thing comes to be of such a quantity or quality or in such a relation, time, or place, a subject is always presupposed, since substance alone is not predicated of another subject, but everything else of substance.

But that substances too, and anything else that can be said 'to be' without qualification, come to be from some substratum, will appear on examination. For we find in every case something that underlies from which proceeds that which comes to be; for instance, animals and plants from seed.

Generally things which come to be, come to be in different ways: (1) by change of shape, as a statue; (2) by addition, as things which grow; (3) by taking away, as the Hermes from the stone; (4) by putting together, as a house; (5) by alteration, as things which 'turn' in respect of their material substance.

It is plain that these are all cases of coming to be from a substratum.

Thus, clearly, from what has been said, whatever comes to be is always complex. There is, on the one hand, (*a*) something which comes into existence, and again (*b*) something which becomes that – the latter (*b*) in two senses, either the subject or the opposite. By the 'opposite' I mean the 'unmusical', by the 'subject' 'man', and similarly I call the absence of shape or form or order the 'opposite', and the bronze or stone or gold the 'subject'.

Plainly then, if there are conditions and principles which constitute natural objects and from which they primarily are or have come to be – have come to be, I mean, what each is said to be in its essential nature, not what each is in respect of a concomitant attribute – plainly, I say, everything comes to be from both subject and form. For 'musical man' is composed (in a way) of 'man' and 'musical': you can analyse it into the definitions of its elements. It is clear then that what comes to be will come to be from these elements.

Now the subject is one numerically, though it is two in form. (For it is the man, the gold – the 'matter' generally – that is counted, for it is more of the nature of a 'this', and what comes to be does not come from it in virtue of a concomitant attribute; the privation, on the other hand, and the contrary *are* incidental in the process.) And the positive form is one – the order, the acquired art of music, or any similar predicate.

There is a sense, therefore, in which we must declare the principles to be two, and a sense in which they are three; a sense in which the contraries are the principles – say for example the musical and the unmusical, the hot and the cold, the tuned and the untuned – and a sense in which they are not, since it is impossible for the contraries to be acted on by each other. But this difficulty also is solved by the fact that the substratum is different from the contraries, for it is itself not a contrary. The principles therefore are, in a way, not more in number than the contraries, but as it were two, nor yet precisely two, since there is a difference of essential nature, but three. For 'to be man' is different from 'to be unmusical', and 'to be unformed' from 'to be bronze'.

We have now stated the number of the principles of natural objects which are subject to generation, and how the number is reached: and it is clear that there must be a substratum for the contraries, and that the contraries must be two. (Yet in another way of putting it this is not necessary, as one of the contraries will serve to effect the change by its successive absence and presence.)

The underlying nature is an object of scientific knowledge, by an analogy. For as the bronze is to the statue, the wood to the bed, or the matter and the formless before receiving form to any thing which has form, so is the underlying nature to substance, i.e. the 'this' or existent.

This then is one principle (though not one or existent in the same sense as the 'this'), and the definition was one as we agreed; then further there is its contrary, the privation. In what sense these are two, and in what sense more, has been stated above. Briefly, we explained first that only the contraries were principles, and later that a substratum was indispensable, and that the principles were three; our last statement has elucidated the difference between the contraries, the mutual relation of the principles, and the nature of the substratum. Whether the form or the substratum is the essential nature of a physical object is not yet clear. But that the principles are three, and in what sense, and the way in which each is a principle, is clear.

So much then for the question of the number and the nature of the principles.

The True Opinion Removes the Difficulty Felt by the Early Philosophers

We will now proceed to show that the difficulty of the early thinkers, as well as our own, is solved in this way alone.

The first of those who studied science were misled in their search for truth and the nature of things by their inexperience, which as it were thrust them into another path.

So they say that none of the things that are either comes to be or passes out of existence, because what comes to be must do so either from what is or from what is not, both of which are impossible. For what is cannot come to be (because it is already), and from what is not nothing could have come to be (because something must be present as a substratum). So too they exaggerated the consequence of this, and went so far as to deny even the *existence* of a plurality of things, maintaining that only Being itself is. Such then was their opinion, and such the reason for its adoption.

Our explanation on the other hand is that the phrases 'something comes to be from what is or from what is not', 'what is not or what is does something or has something done to it or becomes some particular thing', are to be taken (in the first way of putting our explanation) in the same sense as 'a doctor does something or has something done to him', 'is or becomes something from being a doctor'. These expressions may be taken in two senses, and so too, clearly, may 'from being', and 'being acts or is acted on'. A doctor builds a house, not *qua* doctor, but *qua* housebuilder, and turns grey, not *qua* doctor, but *qua* darkhaired. On the other hand he doctors or fails to doctor *qua* doctor. But we are using words most appropriately when we say that a doctor does something or undergoes something, or becomes something from being a doctor, if he does, undergoes, or becomes *qua* doctor. Clearly then also 'to come to be so-and-so from not-being' means '*qua* not-being'.

It was through failure to make this distinction that those thinkers gave the matter up, and through this error that they went so much farther astray as to suppose that nothing else comes to be or exists apart from Being itself, thus doing away with all becoming.

We ourselves are in agreement with them in holding that nothing can be said without qualification to come from what is not. But nevertheless we maintain that a thing may 'come to be from what is not' – that is, in a qualified sense. For a thing comes to be from the privation, which in its own nature is not-being – this not surviving as a constituent of the result. Yet this causes surprise, and it is thought impossible that something should come to be in the way described from what is not.

In the same way we maintain that nothing comes to be from being, and that being does not come to be except in a qualified sense. In that way, however, it does, just as animal might come to be from animal, and an animal of a certain kind from an animal of a certain kind. Thus, suppose a dog to come to be from a horse. The dog would then, it is true, come to be from animal (as well as from an animal of a certain kind) but not as *animal*, for that is already there. But if anything is to become an animal, *not* in a qualified sense, it will not be from animal: and if being, not from being – nor from not-being either, for it has been explained that by 'from not-being' we mean from not-being *qua* not-being.

Note further that we do not subvert the principle that everything either is or is not.

This then is one way of solving the difficulty. Another consists in pointing out that the same things can be explained in terms of potentiality and actuality. But this has been done with greater precision elsewhere.

So, as we said, the difficulties which constrain people to deny the existence of some of the things we mentioned are now solved. For it was this reason which also caused some of the earlier thinkers to turn so far aside from the road which leads to coming to be and passing away and change generally. If they had come in sight of this nature, all their ignorance would have been dispelled.

Further Reflections on the First Principles of Nature

Others, indeed, have apprehended the nature in question, but not adequately.

In the first place they allow that a thing may come to be without qualification from not-being, accepting on this point the statement of Parmenides. Secondly, they think that if the substratum is one numerically, it must have also only a single potentiality – which is a very different thing.

Now we distinguish matter and privation, and hold that one of these, namely the matter, is not-being only in virtue of an attribute which it has, while the privation in its own nature is not-being; and that the matter is nearly, in a sense *is*, substance, while the privation in no sense is. They, on the other hand, identify their Great and Small alike with not-being, and that whether they are taken together as one or separately. Their triad is therefore of quite a different kind from ours. For they got so far as to see that there must be some underlying nature, but they make it one – for even if one philosopher makes a dyad of it, which he calls Great and Small, the effect is the same, for he overlooked the other nature. For the one which persists is a joint cause, with the form, of what comes to be – a mother, as it were. But the negative part of the contrariety may often seem, if you concentrate your attention on it as an evil agent, not to exist at all.

For admitting with them that there is something divine, good, and desirable, we hold that there are two other principles, the one contrary to it, the other such as of its own nature to desire and yearn for it. But the consequence of their view is that the contrary desires its own extinction. Yet the form cannot desire itself, for it is not defective; nor can the contrary desire it, for contraries are mutually destructive. The truth is that what desires the form is matter, as the female desires the male and the ugly the beautiful – only the ugly or the female not *per se* but *per accidens*.

The matter comes to be and ceases to be in one sense, while in another it does not. As that which contains the privation, it ceases to be in its own nature, for what ceases to be – the privation – is contained within it. But as potentiality it does not cease to be in its own nature, but is necessarily outside the sphere of becoming and ceasing to be. For if it came to be, something must have existed as a primary substratum from which it should come and which should persist in it; but this is its own special nature, so that it will be before coming to be. (For my definition of matter is just this – the primary substratum of each thing, from which it comes to be without qualification, and which persists in the result.) And if it ceases to be it will pass into that at the last, so it will have ceased to be before ceasing to be.

The accurate determination of the first principle in respect of form, whether it is one or many and what it is or what they are, is the province of the primary type of science; so these questions may stand over till then. But of the natural, i.e. perishable, forms we shall speak in the expositions which follow.

The above, then, may be taken as sufficient to establish that there are principles and what they are and how many there are. Now let us make a fresh start and proceed.

Book II

Nature and the Natural

Of things that exist, some exist by nature, some from other causes.

'By nature' the animals and their parts exist, and the plants and the simple bodies (earth, fire, air, water) – for we say that these and the like exist 'by nature'.

All the things mentioned present a feature in which they differ from things which are *not* constituted by nature. Each of them has *within itself* a principle of motion and of stationariness (in respect of place, or of growth and decrease, or by way of alteration). On the other hand, a bed and a coat and anything else of that sort, *qua* receiving these designations – i.e. in so far as they are products of art – have no innate impulse to change. But in so far as they happen to be composed of stone or of earth or of a mixture of the two, they *do* have such an impulse, and just to that extent – which seems to indicate that *nature is a source or cause of being moved and of being at rest in that to which it belongs primarily,* in virtue of itself and not in virtue of a concomitant attribute.

I say 'not in virtue of a concomitant attribute', because (for instance) a man who is a doctor might cure himself. Nevertheless it is not in so far as he is a patient that he possesses the art of medicine: it merely has happened that the same man is doctor and patient – and that is why these attributes are not always found together. So it is with all other artificial products. None of them has in itself the source of its own production. But while in some cases (for instance houses and the other products of manual labour) that principle is in something else external to the thing, in others – those which may cause a change in themselves in virtue of a concomitant attribute – it lies in the things themselves (but not in virtue of what they are).

'Nature' then is what has been stated. Things 'have a nature' which have a principle of this kind. Each of them is a substance; for it is a subject, and nature always implies a subject in which it inheres.

The term 'according to nature' is applied to all these things and also to the attributes which belong to them in virtue of what they are, for instance the property of fire to be carried upwards – which is not a 'nature' nor 'has a nature' but is 'by nature' or 'according to nature'.

What nature is, then, and the meaning of the terms 'by nature' and 'according to nature', has been stated. *That* nature exists, it would be absurd to try to prove; for it is obvious that there are many things of this kind, and to prove what is obvious by what is not is the mark of a man who is unable to distinguish what is self-evident from what is not. (This state of mind is clearly possible. A man blind from birth might reason about colours. Presumably therefore such persons must be talking about words without any thought to correspond.)

Some identify the nature or substance of a natural object with that immediate constituent of it which taken by itself is without arrangement, e.g. the wood is the 'nature' of the bed, and the bronze the 'nature' of the statue.

As an indication of this Antiphon points out that if you planted a bed and the rotting wood acquired the power of sending up a shoot, it would not be a bed that would come up, but *wood* – which shows that the arrangement in accordance with the rules of the art is merely an incidental attribute, whereas the real nature is the other, which, further, persists continuously through the process of making.

But if the material of each of these objects has itself the same relation to something else, say bronze (or gold), to water, bones (or wood) to earth and so on, *that* (they say) would be their nature and essence. Consequently some assert earth, others fire or air or water or some or all of these, to be the nature of the things that are. For whatever any one of them supposed to have this character – whether one thing or more than one thing – this or these he declared to be the whole of substance, all else being its affections, states, or dispositions. Every such thing they held to be eternal (for it could not pass into anything else), but other things to come into being and cease to be times without number.

This then is one account of 'nature', namely that it is the immediate material substratum of things which have in themselves a principle of motion or change.

Another account is that 'nature' is the shape or form which is specified in the definition of the thing.

For the word 'nature' is applied to what is according to nature and the natural in the same way as 'art' is applied to what is artistic or a work of art. We should not say in the latter case that there is anything artistic about a thing, if it is a bed only potentially, not yet having the form of a bed; nor should we call it a work of art. The same is true of natural compounds. What is potentially flesh or bone has not yet its own 'nature', and does not exist 'by nature', until it receives the form specified in the definition, which we name in defining what flesh or bone is. Thus in the second sense of 'nature' it would be the shape or form (not separable except in statement) of things which have in themselves a source of motion. (The combination of the two, e.g. man, is not 'nature' but 'by nature' or 'natural'.)

The form indeed is 'nature' rather than the matter; for a thing is more properly said to be what it is when it has attained to fulfilment than when it exists potentially. Again man is born from man, but not bed from bed. That is why people say that the figure is not the nature of a bed, but the wood is – if the bed sprouted not a bed but wood would come up. But even if the figure *is* art, then on the same principle the shape of man is his nature. For man is born from man.

We also speak of a thing's nature as being exhibited in the process of growth by which its nature is attained. The 'nature' in this sense is not like 'doctoring', which leads not to the art of doctoring but to health. Doctoring must start from the art, not lead to it. But it is not in this way that nature (in the one sense) is related to nature (in the other). What grows *qua* growing grows from something into something. Into what then does it grow? Not into that from which it arose but into that to which it tends. The shape then is nature.

'Shape' and 'nature', it should be added, are used in two senses. For the privation too is in a way form. But whether in unqualified coming to be there is privation, i.e. a contrary to what comes to be, we must consider later.

Distinction of the Natural Philosopher from the Mathematician and the Metaphysician

We have distinguished, then, the different ways in which the term 'nature' is used.

The next point to consider is how the mathematician differs from the physicist. Obviously physical bodies contain surfaces and volumes, lines and points, and these are the subject-matter of mathematics.

Further, is astronomy different from physics or a department of it? It seems absurd that the physicist should be supposed to know the nature of sun or moon, but not to know any of their essential attributes, particularly as the writers on physics obviously do discuss their shape also and whether the earth and the world are spherical or not.

Now the mathematician, though he too treats of these things, nevertheless does not treat of them as the limits of a physical body; nor does he consider the attributes indicated as the attributes of such bodies. That is why he separates them; for in thought they are separable from motion, and it makes no difference, nor does any falsity result, if they are separated. The holders of the theory of Forms do the same, though they are not aware of it; for they separate the objects of physics, which are less separable than those of mathematics. This becomes plain if one tries to state in each of the two cases the definitions of the things and of their attributes. 'Odd' and 'even', 'straight' and 'curved', and likewise 'number', 'line', and 'figure', do not involve motion; not so 'flesh' and 'bone' and 'man' – *these* are defined like 'snub nose', not like 'curved'.

Similar evidence is supplied by the more physical of the branches of mathematics, such as optics, harmonics, and astronomy. These are in a way the converse of geometry. While geometry investigates physical lines but not *qua* physical, optics investigates mathematical lines, but *qua* physical, not *qua* mathematical.

Since 'nature' has two senses, the form and the matter, we must investigate its objects as we would the essence of snubness. That is, such things are neither independent of matter nor can be defined in terms of matter only. Here too indeed one might raise a difficulty. Since there are two natures, with which is the physicist concerned? Or should he investigate the combination of the two? But if the combination of the two, then also each severally. Does it belong then to the same or to different sciences to know each severally?

If we look at the ancients, physics would seem to be concerned with the *matter*. (It was only very slightly that Empedocles and Democritus touched on the forms and the essence.)

But if on the other hand art imitates nature, and it is the part of the same discipline to know the form and the matter up to a point (e.g. the doctor has a knowledge of health and also of bile and phlegm, in which health is realized, and the builder both of the form of the house and of the matter, namely that it is bricks and beams, and so forth): if this is so, it would be the part of physics also to know nature in both its senses.

Again, 'that for the sake of which', or the end, belongs to the same department of knowledge as the means. But the nature is the end or 'that for the sake of which'. For if a thing undergoes a continuous change and there is a stage which is last, this stage is the end or 'that for the sake of which'. (That is why the poet was carried away into making an absurd statement when he said, 'He has the end for the sake of which he was born'. For not every stage that is last claims to be an end, but only that which is best.)

For the arts make their material (some simply 'make' it, others make it serviceable), and we use everything as if it was there for our sake. (We also are in a sense an end. 'That for the sake of which' has two senses: the distinction is made in our work *On Philosophy*.) The arts, therefore, which govern the matter and have knowledge are two, namely the art which uses the product and the art which directs the production of it. That is why the using art also is in a sense directive; but it differs in that it knows the form, whereas the art which is directive as being concerned with production knows the matter. For the helmsman knows and prescribes what sort of form a helm should have, the other from what wood it should be made and by means of what operations. In the products of art, however, we make the material with a view to the function, whereas in the products of nature the matter is there all along.

Again, matter is a relative term: to each form there corresponds a special matter. How far then must the physicist know the form or essence? Up to a point, perhaps, as the doctor must know sinew or the smith bronze (i.e. until he understands the purpose of each): and the physicist is concerned only with things whose forms are separable indeed, but do not exist apart from matter. Man is begotten by man and by the sun as well. The mode of existence and essence of the separable it is the business of the primary type of philosophy to define.

The Conditions of Change

The Essential Conditions

Now that we have established these distinctions, we must proceed to consider causes, their character and number. Knowledge is the object of our inquiry, and men do not think they know a thing till they have grasped the 'why' of it (which is to grasp its primary cause). So clearly we too must do this as regards both coming to be and passing away and every kind of physical change, in order that, knowing their principles, we may try to refer to these principles each of our problems.

In one sense, then, (1) that out of which a thing comes to be and which persists, is called 'cause', e.g. the bronze of the statue, the silver of the bowl, and the genera of which the bronze and the silver are species.

In another sense (2) the form or the archetype, i.e. the statement of the essence, and its genera, are called 'causes' (e.g. of the octave the relation of 2:1, and generally number), and the parts in the definition.

Again (3) the primary source of the change or coming to rest; e.g. the man who gave advice is a cause, the father is cause of the child, and generally what makes of what is made and what causes change of what is changed.

Again (4) in the sense of end or 'that for the sake of which' a thing is done, e.g. health is the cause of walking about. ('Why is he walking about?' we say. 'To be healthy', and, having said that, we think we have assigned the cause.) The same is true also of all the intermediate steps which are brought about through the action of something else as means towards the end, e.g. reduction of flesh, purging, drugs, or surgical instruments are means towards health. All these things are 'for the sake of' the end, though they differ from one another in that some are activities, others instruments.

This then perhaps exhausts the number of ways in which the term 'cause' is used.

As the word has several senses, it follows that there are several causes of the same thing (not merely in virtue of a concomitant attribute), e.g. both the art of the sculptors and the bronze are causes of the statue. These are causes of the statue *qua* statue, not in virtue of anything else that it may be – only not in the same way, the one being the material cause, the other the cause whence the motion comes. Some things cause each other reciprocally, e.g. hard work causes fitness and *vice versa*, but again not in the same way, but the one as end, the other as the origin of change. Further the same thing is the cause of contrary results. For that which by its presence brings about one result is sometimes blamed for bringing about the contrary by its absence. Thus we ascribe the wreck of a ship to the absence of the pilot whose presence was the cause of its safety.

All the causes now mentioned fall into four familiar divisions. The letters are the causes of syllables, the material of artificial products, fire, &c., of bodies, the parts of the whole, and the premises of the conclusion, in the sense of 'that from which'. Of these pairs the one set are causes in the sense of substratum, e.g. the parts, the other set in the sense of essence the whole and the combination and the form. But the seed and the doctor and the adviser, and generally the maker, are all sources whence the change or stationariness originates, while the others are causes in the sense of the end or the good of the rest; for 'that for the sake of which' means what is best and the end of the things that lead up to it. (Whether we say the 'good itself' or the 'apparent good' makes no difference.)

Such then is the number and nature of the kinds of cause.

Now the modes of causation are many, though when brought under heads they too can be reduced in number. For 'cause' is used in many senses and even within the same kind one may be prior to another (e.g. the doctor and the expert are causes of health, the relation 2:1 and number of the octave), and always what is inclusive to what is particular. Another mode of causation is the incidental and its genera, e.g. in one way 'Polyclitus', in another 'sculptor' is the cause of a statue, because 'being Polyclitus' and 'sculptor' are incidentally conjoined. Also the classes in which the incidental attribute is included; thus 'a man' could be said to be the cause of a statue or, generally, 'a living creature'. An incidental attribute too may be more or less remote, e.g. suppose that 'a pale man' or 'a musical man' were said to be the cause of the statue.

All causes, both proper and incidental, may be spoken of either as potential or as actual; e.g. the cause of a house being built is either 'house-builder' or 'house-builder building'.

Similar distinctions can be made in the things of which the causes are causes, e.g. of 'this statue' or of 'statue' or of 'image' generally, of 'this bronze' or of 'bronze' or of 'material' generally. So too with the incidental attributes. Again we may use a complex expression for either and say, e.g., neither 'Polyclitus' nor 'sculptor' but 'Polyclitus, sculptor'.

All these various uses, however, come to six in number, under each of which again the usage is twofold. Cause means either what is particular or a genus, or an incidental attribute or a genus of that, and these either as a complex or each by itself;

and all six either as actual or as potential. The difference is this much, that causes which are actually at work and particular exist and cease to exist simultaneously with their effect, e.g. this healing person with this being-healed person and that housebuilding man with that being-built house; but this is not always true of potential causes – the house and the housebuilder do not pass away simultaneously.

In investigating the cause of each thing it is always necessary to seek what is most precise (as also in other things): thus man builds because he is a builder, and a builder builds in virtue of his art of building. This last cause then is prior: and so generally.

Further, generic effects should be assigned to generic causes, particular effects to particular causes, e.g. statue to sculptor, this statue to this sculptor; and powers are relative to possible effects, actually operating causes to things which are actually being effected.

This must suffice for our account of the number of causes and the modes of causation.

The Opinions of Others About Chance and Spontaneity

But chance also and spontaneity are reckoned among causes: many things are said both to be and to come to be as a result of chance and spontaneity. We must inquire therefore in what manner chance and spontaneity are present among the causes enumerated, and whether they are the same or different, and generally what chance and spontaneity are.

Some people even question whether they are real or not. They say that nothing happens by chance, but that everything which we ascribe to chance or spontaneity has some definite cause, e.g. coming 'by chance' into the market and finding there a man whom one wanted but did not expect to meet is due to one's wish to go and buy in the market. Similarly in other cases of chance it is always possible, they maintain, to find something which is the cause; but not chance, for if chance were real, it would seem strange indeed, and the question might be raised, why on earth none of the wise men of old in speaking of the causes of generation and decay took account of chance; whence it would seem that they too did not believe that anything is by chance. But there is a further circumstance that is surprising. Many things both come to be and are by chance and spontaneity, and although all know that each of them can be ascribed to some cause (as the old argument said which denied chance), nevertheless they speak of some of these things as happening by chance and others not. For this reason also they ought to have at least referred to the matter in some way or other.

Certainly the early physicists found no place for chance among the causes which they recognized – love, strife, mind, fire, or the like. This is strange, whether they supposed that there is no such thing as chance or whether they thought there is but omitted to mention it – and that too when they sometimes used it, as Empedocles does when he says that the air is not always separated into the highest region, but 'as it may chance'. At any rate he says in his cosmogony that 'it happened to run that way at that time, but it often ran otherwise.' He tells us also that most of the parts of animals came to be by chance.

There are some too who ascribe this heavenly sphere and all the worlds to spontaneity. They say that the vortex arose spontaneously, i.e. the motion that separated and arranged in its present order all that exists. This statement might well cause surprise. For they are asserting that chance is not responsible for the existence or generation of animals and plants, nature or mind or something of the kind being the cause of them (for it is not any chance thing that comes from a given seed but an olive from one kind and a man from another); and yet at the same time they assert that the heavenly sphere and the divinest of visible things arose spontaneously, having no such cause as is assigned to animals and plants. Yet if this is so, it is a fact which deserves to be dwelt upon, and something might well have been said about it. For besides the other absurdities of the statement, it is the more absurd that people should make it when they see nothing coming to be spontaneously in the heavens, but much happening by chance among the things which as they say are not due to chance; whereas we should have expected exactly the opposite.

Others there are who, indeed, believe that chance is a cause, but that it is inscrutable to human intelligence, as being a divine thing and full of mystery.

Thus we must inquire what chance and spontaneity are, whether they are the same or different, and how they fit into our division of causes.

Do Chance and Spontaneity Exist? What Is Chance and What Are Its Characteristics?

First then we observe that some things always come to pass in the same way, and others for the most part. It is clearly of neither of these that chance is said to be the cause, nor can the 'effect of chance' be identified with any of the things that come to pass by necessity and always, or for the most part. But as there is a third class of events besides these two – events which all say are 'by chance' – it is plain that there is such a thing as chance and spontaneity; for we know that things of this kind are due to chance and that things due to chance are of this kind.

But, secondly, some events are for the sake of something, others not. Again, some of the former class are in accordance with deliberate intention, others not, but both are in the class of things which are for the sake of something. Hence it is clear that even among the things which are outside the necessary and the normal, there are some in connexion with which the phrase 'for the sake of something' is applicable. (Events that are for the sake of something include whatever may be done as a result of thought or of nature.) Things of this kind, then, when they come to pass incidentally are said to be 'by chance'. For just as a thing is something either in virtue of itself or incidentally, so may it be a cause. For instance, the housebuilding faculty is in virtue of itself the cause of a house, whereas the pale or the musical is the incidental cause. That which is *per se* cause of the effect is determinate, but the incidental cause is indeterminable, for the possible attributes of an individual are innumerable. To resume then; when a thing of this kind comes to pass among events which are for the sake of something, it is said to be spontaneous or by chance. (The distinction between the two must be made later – for the present it is sufficient if it is plain that both are in the sphere of things done for the sake of something.)

Example: A man is engaged in collecting subscriptions for a feast. He would have gone to such and such a place for the purpose of getting the money, if he had known. He actually went there for another purpose, and it was only incidentally that he got his money by going there; and this was not due to the fact that he went there as a rule or necessarily, nor is the end effected (getting the money) a cause present in himself – it belongs to the class of things that are intentional and the result of intelligent deliberation. It is when these conditions are satisfied that the man is said to have gone 'by chance'. If he had gone of deliberate purpose and for the sake of this – if he always or normally went there when he was collecting payments – he would not be said to have gone 'by chance'.

It is clear then that chance is an incidental cause in the sphere of those actions for the sake of something which involve purpose. Intelligent reflection, then, and chance are in the same sphere, for purpose implies intelligent reflection. It is necessary, no doubt, that the causes of what comes to pass by chance be indefinite; and that is why chance is supposed to belong to the class of the indefinite and to be inscrutable to man, and why it might be thought that, in a way, nothing occurs by chance. For all these statements are correct, because they are well grounded. Things do, in a way, occur by chance, for

73

they occur incidentally and chance is an *incidental cause*. But strictly it is not the *cause* – without qualification – of anything; for instance, a housebuilder is the cause of a house; incidentally, a fluteplayer may be so.

And the causes of the man's coming and getting the money (when he did not come for the sake of that) are innumerable. He may have wished to see somebody or been following somebody or avoiding somebody, or may have gone to see a spectacle. Thus to say that chance is a thing contrary to rule is correct. For 'rule' applies to what is always true or true for the most part, whereas chance belongs to a third type of event. Hence, to conclude, since causes of this kind are indefinite, chance too is indefinite. (Yet in some cases one might raise the question whether *any* incidental fact might be the cause of the chance occurrence, e.g. of health, the fresh air or the sun's heat may be the cause, but having had one's hair cut *cannot*; for some incidental causes are more relevant to the effect than others.)

Chance or fortune is called 'good' when the result is good, 'evil' when it is evil. The terms 'good fortune' and 'ill fortune' are used when either result is of considerable magnitude. Thus one who comes within an ace of some great evil or great good is said to be fortunate or unfortunate. The mind affirms the presence of the attribute, ignoring the hair's breadth of difference. Further, it is with reason that good fortune is regarded as unstable; for chance is unstable, as none of the things which result from it can be invariable or normal.

Both are then, as I have said, incidental causes – both chance and spontaneity – in the sphere of things which are capable of coming to pass not necessarily, nor normally, and with reference to such of these as might come to pass for the sake of something.

Distinction Between Chance and Spontaneity, and Between Both and the Essential Conditions of Change

They differ in that 'spontaneity' is the wider term. Every result of chance is from what is spontaneous, but not everything that is from what is spontaneous is from chance.

Chance and what results from chance are appropriate to agents that are capable of good fortune and of moral action generally. Therefore necessarily chance is in the sphere of moral actions. This is indicated by the fact that good fortune is thought to be the same, or nearly the same, as happiness, and happiness to be a kind of moral action, since it is well-doing. Hence what is not capable of moral action cannot do anything by chance. Thus an inanimate thing or a lower animal or a child cannot do anything by chance, because it is incapable of deliberate intention; nor can 'good fortune' or 'ill fortune' be ascribed to them, except metaphorically, as Protarchus, for example, said that the stones of which altars are made are fortunate because they are held in honour, while their fellows are trodden under foot. Even these things, however, can in a way be affected by chance, when one who is dealing with them does something to them by chance, but not otherwise.

The spontaneous on the other hand is found both in the lower animals and in many inanimate objects. We say, for example, that the horse came 'spontaneously', because, though his coming saved him, he did not come for the sake of safety. Again, the tripod fell 'of itself', because, though when it fell it stood on its feet so as to serve for a seat, it did not fall for the sake of that.

Hence it is clear that events which (1) belong to the general class of things that may come to pass for the sake of something, (2) do not come to pass for the sake of what actually results, and (3) have an external cause, may be described by the phrase 'from spontaneity'. These 'spontaneous' events are said to be 'from chance' if they have the further characteristics of being the objects of deliberate intention and due to agents capable of that mode of action. This is indicated by the phrase 'in vain', which is used when A, which is for the sake of B, does not result in B. For instance, taking a walk is for the sake of evacuation of the bowels; if this does not follow after walking, we say that we have walked 'in vain' and that the walking was 'vain'. This implies that what is naturally the means to an end is 'in vain', when it does not effect the end towards which it was the natural means – for it would be absurd for a man to say that he had bathed in vain because the sun was not eclipsed, since the one was not done with a view to the other. Thus the spontaneous is even according to its derivation the case in which the thing itself happens in vain. The stone that struck the man did not fall for the purpose of striking him; therefore it fell spontaneously, because it might have fallen by the action of an agent and for the purpose of striking. The difference between spontaneity and what results by chance is greatest in things that come to be by nature; for when anything comes to be contrary to nature, we do not say that it came to be by chance, but by spontaneity. Yet strictly this too is different from the spontaneous proper; for the cause of the latter is external, that of the former internal.

We have now explained what chance is and what spontaneity is, and in what they differ from each other. Both belong to the mode of causation 'source of change', for either some natural or some intelligent agent is always the cause; but in this sort of causation the number of possible causes is infinite.

Spontaneity and chance are causes of effects which, though they might result from intelligence or nature, have in fact been caused by something *incidentally*. Now since nothing which is incidental is prior to what is *per se*, it is clear that no incidental cause can be prior to a cause *per se*. Spontaneity and chance, therefore, are posterior to intelligence and nature. Hence, however true it may be that the heavens are due to spontaneity, it will still be true that intelligence and nature will be prior causes of this All and of many things in it besides.

Proof in Natural Philosophy

The Physicist Demonstrates by Means of the Four Conditions of Change

It is clear then that there are causes, and that the number of them is what we have stated. The number is the same as that of the things comprehended under the question 'why'. The 'why' is referred ultimately either (1) in things which do not involve motion, e.g. in mathematics, to the 'what' (to the definition of 'straight line' or 'commensurable', &c.), or (2) to what initiated a motion, e.g. 'why did they go to war?' – 'because there had been a raid'; or (3) we are inquiring 'for the sake of what?' – 'that they may rule'; or (4), in the case of things that come into being, we are looking for the matter. The causes, therefore, are these and so many in number.

Now, the causes being four, it is the business of the physicist to know about them all, and if he refers his problems back to all of them, he will assign the 'why' in the way proper to his science – the matter, the form, the mover, 'that for the sake of which'. The last three often coincide; for the 'what' and 'that for the sake of which' are one, while the primary source of motion is the same in species as these (for man generates man), and so too, in general, are all things which cause movement by being themselves moved; and such as are not of this kind are no longer inside the province of physics, for they cause motion not by possessing motion or a source of motion in themselves, but being themselves incapable of motion. Hence there are three branches of study, one of things which are incapable of motion, the second of things in motion, but indestructible, the third of destructible things.

The question 'why', then, is answered by reference to the matter, to the form, and to the primary moving cause. For in respect of coming to be it is mostly in this last way that causes are investigated – 'what comes to be after what? what was the primary agent or patient?' and so at each step of the series.

Now the principles which cause motion in a physical way are two, of which one is not physical, as it has no principle of motion in itself. Of this kind is whatever causes movement, not being itself moved, such as (1) that which is completely unchangeable, the primary reality, and (2) the essence of that which is coming to be, i.e. the form; for this is the end or 'that for the sake of which'. Hence since nature is for the sake of something, we must know this cause also. We must explain the 'why' in all the senses of the term, namely, (1) that from this that will necessarily result ('from this' either without qualification or in most cases); (2) that 'this must be so if that is to be so' (as the conclusion presupposes the premisses); (3) that this was the essence of the thing; and (4) because it is better thus (not without qualification, but with reference to the essential nature in each case).

Does Nature Act for an End?

We must explain then (1) that Nature belongs to the class of causes which act for the sake of something; (2) about the necessary and its place in physical problems, for all writers ascribe things to this cause, arguing that since the hot and the cold, &c., are of such and such a kind, therefore certain things *necessarily* are and come to be – and if they mention any other cause (one his 'friendship and strife', another his 'mind'), it is only to touch on it, and then good-bye to it.

A difficulty presents itself: why should not nature work, not for the sake of something, nor because it is better so, but just as the sky rains, not in order to make the corn grow, but of necessity? What is drawn up must cool, and what has been cooled must become water and descend, the result of this being that the corn grows. Similarly if a man's crop is spoiled on the threshing-floor, the rain did not fall for the sake of this – in order that the crop might be spoiled – but that result just followed. Why then should it not be the same with the parts in nature, e.g. that our teeth should come up *of necessity* – the front teeth sharp, fitted for tearing, the molars broad and useful for grinding down the food – since they did not arise for this end, but it was merely a coincident result; and so with all other parts in which we suppose that there is purpose? Wherever then all the parts came about just what they would have been if they had come to be for an end, such things survived, being organized spontaneously in a fitting way; whereas those which grew otherwise perished and continue to perish, as Empedocles says his 'man-faced ox-progeny' did.

Such are the arguments (and others of the kind) which may cause difficulty on this point. Yet it is impossible that this should be the true view. For teeth and all other natural things either invariably or normally come about in a given way; but of not one of the results of chance or spontaneity is this true. We do not ascribe to chance or a mere coincidence the frequency of rain in winter, but frequent rain in summer we do; nor heat in the dog-days, but only if we have it in winter. If then, it is agreed that things are either the result of coincidence or for an end, and these cannot be the result of coincidence or spontaneity, it follows that they must be for an end; and that such things are all due to nature even the champions of the theory which is before us would agree. Therefore action for an end is present in things which come to be and are by nature.

Further, where a series has a completion, all the preceding steps are for the sake of that. Now surely as in intelligent action, so in nature; and as in nature, so it is in each action, if nothing interferes. Now intelligent action is for the sake of an end; therefore the nature of things also is so. Thus if a house, e.g. had been a thing made by nature, it would have been made in the same way as it is now by art; and if things made by nature were made also by art, they would come to be in the same way as by nature. Each step then in the series is for the sake of the next; and generally art partly completes what nature cannot bring to a finish, and partly imitates her. If, therefore, artificial products are for the sake of an end, so clearly also are natural products. The relation of the later to the earlier terms of the series is the same in both.

This is most obvious in the animals other than man: they make things neither by art nor after inquiry or deliberation. Wherefore people discuss whether it is by intelligence or by some other faculty that these creatures work – spiders, ants, and the like. By gradual advance in this direction we come to see clearly that in plants too that is produced which is conducive to the end – leaves, e.g. grow to provide shade for the fruit. If then it is both by nature and for an end that the swallow makes its nest and the spider its web, and plants grow leaves for the sake of the fruit and send their roots down (not up) for the sake of nourishment, it is plain that this kind of cause is operative in things which come to be and are by nature. And since 'nature' means two things, the matter and the form, of which the latter is the end, and since all the rest is for the sake of the end, the form must be the cause in the sense of 'that for the sake of which'.

Now mistakes come to pass even in the operations of art: the grammarian makes a mistake in writing and the doctor pours out the wrong dose. Hence clearly mistakes are possible in the operations of nature also. If then in art there are cases in which what is rightly produced serves a purpose, and if where mistakes occur there was a purpose in what was attempted, only it was not attained, so must it be also in natural products, and monstrosities will be failures in the purposive effort. Thus in the original combinations the 'ox-progeny' if they failed to reach a determinate end must have arisen through the corruption of some principle corresponding to what is now the seed.

Further, seed must have come into being first, and not straightway the animals: the words 'whole-natured first...' must have meant seed.

Again, in plants too we find the relation of means to end, though the degree of organization is less. Were there then in plants also 'olive-headed vine-progeny', like the 'manheaded ox-progeny', or not? An absurd suggestion; yet there must have been, if there were such things among animals.

Moreover, among the seeds anything must have come to be at random. But the person who asserts this entirely does away with 'nature' and what exists 'by nature'. For those things are natural which, by a continuous movement originated from an internal principle, arrive at some completion: the same completion is not reached from every principle; nor any chance completion, but always the tendency in each is towards the same end, if there is no impediment.

The end and the means towards it may come about by chance. We say, for instance, that a stranger has come by chance, paid the ransom, and gone away, when he does so as if he had come for that purpose, though it was not for that that he came. This is incidental, for chance is an incidental cause, as I remarked before. But when an event takes place always or for the most part, it is not incidental or by chance. In natural products the sequence is invariable, if there is no impediment.

It is absurd to suppose that purpose is not present because we do not observe the agent deliberating. Art does not deliberate. If the ship-building art were in the wood, it would produce the same results *by nature*. If, therefore, purpose is present in art, it is present also in nature. The best illustration is a doctor doctoring himself: nature is like that.

It is plain then that nature is a cause, a cause that operates for a purpose.

The Sense in Which Necessity Is Present in Natural Things

As regards what is 'of necessity', we must ask whether the necessity is 'hypothetical', or 'simple' as well. The current view places what is of necessity in the process of production, just as if one were to suppose that the wall of a house necessarily comes to be because what is heavy is naturally carried downwards and what is light to the top, wherefore the stones and foundations take the lowest place, with earth above because it is lighter, and wood at the top of all as being the lightest. Whereas, though the wall does not come to be *without* these, it is not *due* to these, except as its material cause: it comes to be for the sake of sheltering and guarding certain things. Similarly in all other things which involve production for an end; the product cannot come to be without things which have a necessary nature, but it is not due to these (except as its material); it comes to be for an end. For instance, why is a saw such as it is? To effect so-and-so and for the sake of so-and-so. This end, however, cannot be realized unless the saw is made of iron. It is, therefore, necessary for it to be of iron, if we are to have a saw and perform the operation of sawing. What is necessary then, is necessary *on a hypothesis*; it is not a result necessarily determined by antecedents. Necessity is in the matter, while 'that for the sake of which' is in the definition.

Necessity in mathematics is in a way similar to necessity in things which come to be through the operation of nature. Since a straight line is what it is, it is necessary that the angles of a triangle should equal two right angles. But not conversely; though if the angles are *not* equal to two right angles, then the straight line is not what it is either. But in things which come to be for an end, the reverse is true. If the end is to exist or does exist, that also which precedes it will exist or does exist; otherwise just as there, if the conclusion is not true, the premiss will not be true, so here the end or 'that for the sake of which' will not exist. For this too is itself a starting-point, but of the reasoning, not of the action; while in mathematics the starting-

84

point is the starting-point of the reasoning only, as there is no action. If then there is to be a house, such-and-such things must be made or be there already or exist, or generally the matter relative to the end, bricks and stones if it is a house. But the end is not due to these except as the matter, nor will it come to exist because of them. Yet if they do not exist at all, neither will the house, or the saw – the former in the absence of stones, the latter in the absence of iron – just as in the other case the premisses will not be true, if the angles of the triangle are not equal to two right angles.

The necessary in nature, then, is plainly what we call by the name of matter, and the changes in it. Both causes must be stated by the physicist, but especially the end; for that is the cause of the matter, not *vice versa*; and the end is 'that for the sake of which', and the beginning starts from the definition or essence; as in artificial products, since a house is of such-and-such a kind, certain things must *necessarily* come to be or be there already, or since health is this, these things must necessarily come to be or be there already. Similarly if man is this, then these; if these, then those. Perhaps the necessary is present also in the definition. For if one defines the operation of sawing as being a certain kind of dividing, then this cannot come about unless the saw has teeth of a certain kind; and these cannot be unless it is of iron. For in the definition too there are some parts that are, as it were, its matter.

Book III

Motion

The Nature of Motion

Nature has been defined as a 'principle of motion and change', and it is the subject of our inquiry. We must therefore see that we understand the meaning of 'motion'; for if it were unknown, the meaning of 'nature' too would be unknown.

When we have determined the nature of motion, our next task will be to attack in the same way the terms which are involved in it. Now motion is supposed to belong to the class of things which are *continuous*; and the *infinite* presents itself first in the continuous – that is how it comes about that 'infinite' is often used in definitions of the continuous ('what is infinitely divisible is continuous'). Besides these, *place*, *void*, and *time* are thought to be necessary conditions of motion.

Clearly, then, for these reasons and also because the attributes mentioned are common to, and coextensive with, all the objects of our science, we must first take each of them in hand and discuss it. For the investigation of special attributes comes after that of the common attributes.

To begin then, as we said, with motion.

We may start by distinguishing (1) what exists in a state of fulfilment only, (2) what exists as potential, (3) what exists as potential and also in fulfilment – one being a 'this', another 'so much', a third 'such', and similarly in each of the other modes of the predication of being.

Further, the word 'relative' is used with reference to (1) excess and defect, (2) agent and patient and generally what can move and what can be moved. For 'what can cause movement' is relative to 'what can be moved', and *vice versa*.

Again, there is no such thing as motion *over and above* the things. It is always with respect to substance or to quantity or to quality or to place that what changes changes. But it is impossible, as we assert, to find anything *common* to these which is neither 'this' nor *quantum* nor *quale* nor any of the other predicates. Hence neither will motion and change have reference to something over and above the things mentioned, for there *is* nothing over and above them.

Now each of these belongs to all its subjects in either of two ways: namely (1) substance – the one is positive form, the other privation; (2) in quality, white and black; (3) in quantity, complete and incomplete; (4) in respect of locomotion, upwards and downwards or light and heavy. Hence there are as many types of motion or change as there are meanings of the word 'is'.

We have now before us the distinctions in the various classes of being between what is fully real and what is potential.

Def. *The fulfilment of what exists potentially, in so far as it exists potentially, is motion* – namely, of what is alterable *qua* alterable, *alteration*: of what can be increased and its opposite what can be decreased (there is no common name), *increase* and *decrease*: of what can come to be and can pass away, *coming to be* and *passing away*: of what can be carried along, *locomotion*.

Examples will elucidate this definition of motion. When the buildable, in so far as it is just *that*, is fully real, it is *being built*, and this is build*ing*. Similarly, learning, doctoring, rolling, leaping, ripening, ageing.

The same thing, if it is of a certain kind, can be both potential and fully real, not indeed at the same time or not in the same respect, but e.g. potentially hot and actually cold. Hence at once such things will act and be acted on by one another in many ways: each of them will be capable at the same time of causing alteration and of being altered. Hence, too, what effects motion as a physical agent can be moved: when a thing of this kind causes motion, it is itself also moved. This, indeed, has led some people to suppose that every mover is moved. But this question depends on another set of arguments, and the truth will be made clear later. It is possible for a thing to cause motion, though it is itself incapable of being moved.

It is the fulfilment of what is potential when it is already fully real and operates not as *itself* but as *movable*, that is motion. What I mean by 'as' is this: Bronze is potentially a statue. But it is not the fulfilment of bronze as *bronze* which is motion. For 'to be bronze' and 'to be a certain potentiality' are not the same. If they were identical without qualification, i.e. in *definition*, the fulfilment of bronze as bronze *would* have been motion. But they are not the same, as has been said. (This is obvious in contraries. 'To be capable of health' and 'to be capable of illness' are not the same, for if they were there would be no difference between being ill and being well. Yet the *subject* both of health and of sickness – whether it is humour or blood – is one and the same.)

We can distinguish, then, between the two – just as, to give another example, 'colour' and 'visible' are different – and clearly it is the fulfilment of what is potential as potential that is motion. So this, precisely, is motion.

Further it is evident that motion is an attribute of a thing just *when* it is fully real in this way, and neither before nor after. For each thing of this kind is capable of being at one time actual, at another not. Take for instance the buildable as buildable. The actuality of the buildable as buildable is the process of building. For the actuality of the buildable must be either this or the house. But when there is a house, the buildable is no longer buildable. On the other hand, it is the buildable which is being *built*. The process then of being built must be the kind of actuality required. But building is a kind of motion, and the same account will apply to the other kinds also.

The soundness of this definition is evident both when we consider the accounts of motion that the others have given, and also from the difficulty of defining it otherwise.

One could not easily put motion and change in another genus – this is plain if we consider where some people put it; they identify motion with 'difference' or 'inequality' or 'not being'; but such things are not necessarily moved, whether they are 'different' or 'unequal' or 'non-existent'; nor is change either *to* or from *these* rather than to or from their opposites.

The reason why they put motion into these genera is that it is thought to be something indefinite, and the principles in the second column are indefinite because they are privative: none of them is either 'this' or 'such' or comes under any of the other modes of predication. The reason in turn why motion is thought to be indefinite is that it cannot be classed simply as a potentiality or as an actuality – a thing that is merely *capable* of having a certain size is not undergoing change, nor yet a thing that is *actually* of a certain size, and motion is thought to be a sort of *actuality*, but incomplete, the reason for this view being that the potential whose actuality it is is incomplete. This is why it is hard to grasp what motion is. It is necessary to class it with privation or with potentiality or with sheer actuality, yet none of these seems possible. There remains then the suggested mode of definition, namely that it is a sort of actuality, or actuality of the kind described, hard to grasp, but not incapable of existing.

The mover too is moved, as has been said – every mover, that is, which is capable of motion, and whose immobility is rest – when a thing is subject to motion its immobility is rest. For to act on the movable as such is just to *move* it. But this it does by *contact*, so that at the same time it is also acted on. Hence we can define motion as *the fulfilment of the movable* qua *movable, the cause of the attribute being contact with what can move*, so that the mover is also acted on. The mover or agent will always be the vehicle of a form, either a 'this' or a 'such', which, when it acts, will be the source and cause of the change, e.g. the full-formed man begets man from what is potentially man.

The Mover and the Moved

The solution of the difficulty that is raised about the motion – whether it is in the *movable* – is plain. It is the fulfilment of this potentiality, and by the action of that which has the power of causing motion; and the actuality of that which has the power of causing motion is not other than the actuality of the movable, for it must be the fulfilment of *both*. A thing is capable of causing motion because it *can* do this, it is a mover because it actually *does* it. But it is on the movable that it is capable of acting. Hence there is a single actuality of both alike, just as one to two and two to one are the same interval, and the steep ascent and the steep descent are one – for these are one and the same, although they can be described in different ways. So it is with the mover and the moved.

This view has a dialectical difficulty. Perhaps it is necessary that the actuality of the agent and that of the patient should not be the same. The one is 'agency' and the other 'patiency'; and the outcome and completion of the one is an 'action', that of the other a 'passion'. Since then they are both motions, we may ask: *in* what are they, if they are different? Either (*a*) both are in what is acted on and moved, or (*b*) the agency is in the agent and the patiency in the patient. (If we ought to call the latter also 'agency', the word would be used in two senses.)

Now, in alternative (*b*), the motion will be in the mover, for the same statement will hold of 'mover' and' moved'. Hence either *every* mover will be moved, or, though having motion, it will not be moved.

If on the other hand (*a*) both are in what is moved and acted on – both the agency and the patiency (e.g. both teaching and learning, though they are *two*, in the *learner*), then, first, the actuality of each will not be present in each, and, a second absurdity, a thing will have two motions at the same time. How will there be two alterations of quality in *one* subject towards *one* definite quality? The thing is impossible: the actualization will be one.

But (someone will say) it is contrary to reason to suppose that there should be one identical actualization of two things which are different in kind. Yet there will be, if teaching and learning are the same, and agency and patiency. To teach will be the same as to learn, and to act the same as to be acted on – the teacher will necessarily be learning everything that he teaches, and the agent will be acted on.

One may reply:

(1) It is *not* absurd that the actualization of one thing should be in another. Teaching is the activity of a person who can teach, yet the operation is performed *on* some patient – it is not cut adrift from a subject, but is of *A* on *B*.

(2) There is nothing to prevent two things having one and the same actualization, provided the actualizations are not

described in the same way, but are related as what can act to what is acting.

(3) Nor is it necessary that the teacher should learn, even if to act and to be acted on are one and the same, provided they are not the same in *definition* (as 'raiment' and 'dress'), but are the same merely in the sense in which the road from Thebes to Athens and the road from Athens to Thebes are the same, as has been explained above. For it is not things which are in a way the same that have all their attributes the same, but only such as have the same definition. But indeed it by no means follows from the fact that teaching is the same as learning, that to learn is the same as to teach, any more than it follows from the fact that there is one *distance* between two things which are at a distance from each other, that the two *vectors AB* and *BA* are one and the same. To generalize, teaching is not the same as learning, or agency as patiency, in the full sense, though they belong to the same *subject*, the motion; for the 'actualization of *X* in *Y*' and the 'actualization of *Y* through the action of *X*' differ in *definition*.

What then Motion is, has been stated both generally and particularly. It is not difficult to see how each of its types will be defined – alteration is the fulfilment of the alterable *qua* alterable (or, more scientifically, the fulfilment of what can act and what can be acted on, as such) – generally and again in each particular case, building, healing, &c. A similar definition will apply to each of the other kinds of motion.

The Infinite

Opinions of the Early Philosophers

The science of nature is concerned with spatial magnitudes and motion and time, and each of these at least is necessarily infinite or finite, even if some things dealt with by the science are not, e.g. a quality or a point – it is not necessary perhaps that such things should be put under either head. Hence it is incumbent on the person who specializes in physics to discuss the infinite and to inquire whether there is such a thing or not, and, if there is, *what* it is.

The appropriateness to the science of this problem is clearly indicated. All who have touched on this kind of science in a way worth considering have formulated views about the infinite, and indeed, to a man, make it a principle of things.

(1) Some, as the Pythagoreans and Plato, make the infinite a principle in the sense of a self-subsistent substance, and not as a mere attribute of some other thing. Only the Pythagoreans place the infinite among the objects of sense (they do not regard number as separable from these), and assert that what is outside the heaven is infinite. Plato, on the other hand, holds that there is no body outside (the Forms are not outside, because they are nowhere), yet that the infinite is present not only in the objects of sense but in the Forms also.

Further, the Pythagoreans identify the infinite with the even. For this, they say, when it is cut off and shut in by the odd, provides things with the element of infinity. An indication of this is what happens with numbers. If the gnomons are placed round the one, and without the one, in the one construction the figure that results is always different, in the other it is always the same. But Plato has two infinites, the Great and the Small.

The physicists, on the other hand, all of them, always regard the infinite as an attribute of a substance which is different from it and belongs to the class of the so-called elements – water or air or what is intermediate between them. Those who make them limited in number never make them infinite in amount. But those who make the elements infinite in number, as Anaxagoras and Democritus do, say that the infinite is continuous by contact – compounded of the homogeneous parts according to the one, of the seedmass of the atomic shapes according to the other.

Further, Anaxagoras held that any part is a mixture in the same way as the All, on the ground of the observed fact that anything comes out of anything. For it is probably for this reason that he maintains that once upon a time all things were together. (*This* flesh and *this* bone were together, and so of *any* thing: therefore *all* things: and at the same time too.) For there is a beginning of separation, not only for each thing, but for all. Each thing that comes to be comes to be from a similar body, and there is a coming to be of all things, though not, it is true, at the same time. Hence there must also be an origin of coming to be. One such source there is which he calls Mind, and Mind begins its work of thinking from some starting-point. So necessarily all things must have been together at a certain time, and must have begun to be moved at a certain time.

Democritus, for his part, asserts the contrary, namely that no element arises from another element. Nevertheless for him the common body is a source of all things, differing from part to part in size and in shape.

It is clear then from these considerations that the inquiry concerns the physicist. Nor is it without reason that they all make it a principle or source. We cannot say that the infinite has no effect, and the only effectiveness which we can ascribe to it is that of a principle. Everything is either a source or derived from a source. But there cannot be a source of the infinite or limitless, for that would be a limit of it. Further, as it is a beginning, it is both uncreatable and indestructible. For there must be a point at which what has come to be reaches completion, and also a termination of all passing away. That is why, as we say, there is no principle of *this*, but it is this which is held to be the principle of other things, and to encompass all and to steer all, as those assert who do not recognize, alongside the infinite, other causes, such as Mind or Friendship. Further they identify it with the Divine, for it is 'deathless and imperishable' as Anaximander says, with the majority of the physicists.

Main Arguments for Belief in the Infinite

Belief in the existence of the infinite comes mainly from five considerations:

(1) From the nature of time – for it is infinite.

(2) From the division of magnitudes – for the mathematicians also use the notion of the infinite.

(3) If coming to be and passing away do not give out, it is only because that from which things come to be is infinite.

(4) Because the limited always finds its limit in something, so that there must be no limit, if everything is *always* limited by something different from itself.

(5) Most of all, a reason which is peculiarly appropriate and presents the difficulty that is felt by everybody – not only number but also mathematical magnitudes and what is outside the heaven are supposed to be infinite because they never give out in our *thought*.

The last fact (that what is outside is infinite) leads people to suppose that body also is infinite, and that there is an infinite number of worlds. Why should there be body in one part of the void rather than in another? Grant only that mass is anywhere and it follows that it must be everywhere. Also, if void and place are infinite, there must be infinite body too, for in the case of eternal things what may be must be.

But the problem of the infinite is difficult: many contradictions result whether we suppose it to exist or not to exist. If it exists, we have still to ask *how* it exists; as a substance or as the essential attribute of some entity? Or in neither way, yet none the less is there something which is infinite or some things which are infinitely many?

The problem, however, which specially belongs to the physicist is to investigate whether there is a sensible magnitude which is infinite.

We must begin by distinguishing the various senses in which the term 'infinite' is used.

(1) What is incapable of being gone through, because it is not its nature to be gone through (the sense in which the voice is 'invisible').

(2) What admits of being gone through, the process however having no termination, or (3) what scarcely admits of being gone through.

(4) What naturally admits of being gone through, but is not actually gone through or does not actually reach an end.

Further, everything that is infinite may be so in respect of addition or division or both.

Criticism of the Pythagorean and Platonic Belief in a Separately Existing Infinite

Now it is impossible that the infinite should be a thing which is itself infinite, separable from sensible objects. If the infinite is neither a magnitude nor an aggregate, but is itself a substance and not an attribute, it will be indivisible; for the divisible must be either a magnitude or an aggregate. But if indivisible, then not infinite, except in the sense (1) in which the voice is 'invisible'. But this is not the sense in which it is used by those who say that the infinite exists, nor that in which we are investigating it, namely as (2), 'that which cannot be gone through'. But if the infinite exists as an attribute, it would not be, *qua* infinite, an element in substances, any more than the invisible would be an element of speech, though the voice is invisible.

Further, how can the infinite be itself any thing, unless both number and magnitude, of which it is an essential attribute, exist in that way? If *they* are not substances, *a fortiori* the infinite is not.

It is plain, too, that the infinite cannot be an actual thing and a substance and principle. For any part of it that is taken will be infinite, if it has parts: for 'to be infinite' and 'the infinite' are the same, if it is a substance and not predicated of a subject. Hence it will be either indivisible or divisible into infinites. But the same thing cannot be many infinites. (Yet just as part of air is air, so a part of the infinite would be infinite, if it is supposed to be a substance and principle.) Therefore the infinite must be without parts and indivisible. But this cannot be true of what is infinite in full completion: for it must be a definite quantity.

Suppose then that infinity belongs to substance as an attribute. But, if so, it cannot, as we have said, be described as a principle, but rather that of which it is an attribute – the air or the even number.

Thus the view of those who speak after the manner of the Pythagoreans is absurd. With the same breath they treat the infinite as substance, and divide it into parts.

There Is No Infinite Sensible Body

This discussion, however, involves the more general question whether the infinite can be present in mathematical objects and things which are intelligible and do not have extension, as well as among sensible objects. Our inquiry (as physicists) is limited to its special subject-matter, the objects of sense, and we have to ask whether there is or is not among *them* a body which is infinite in the direction of increase.

We may begin with a dialectical argument and show as follows that there is no such thing.

If 'bounded by a surface' is the definition of body there cannot be an infinite body either intelligible or sensible. Nor can number taken in abstraction be infinite, for number or that which has number is numerable. If then the numerable can be numbered, it would also be possible to go through the infinite.

If, on the other hand, we investigate the question more in accordance with principles appropriate to physics, we are led as follows to the same result.

The infinite body must be either (1) compound, or (2) simple; yet neither alternative is possible.

(1) Compound the infinite body will not be, if the elements are finite in number. For they must be more than one, and the contraries must always balance, and no *one* of them can be infinite. If one of the bodies falls in any degree short of the other in potency – suppose fire is finite in amount while air is infinite and a given quantity of fire exceeds in power the same amount of air in any ratio provided it is numerically definite – the infinite body will obviously prevail over and annihilate the finite body. On the other hand, it is impossible that *each* should be infinite. 'Body' is what has extension in all directions and the infinite is what is boundlessly extended, so that the infinite body would be extended in all directions *ad infinitum.*

Nor (2) can the infinite body be one and simple, whether it is, as some hold, a thing over and above the elements (from which they generate the elements) or is not thus qualified.

(*a*) We must consider the former alternative; for there *are* some people who make this the infinite, and not air or water, in order that the other elements may not be annihilated by the element which is infinite. They have contrariety with each other – air is cold, water moist, fire hot; if one were infinite, the others by now would have ceased to be. As it is, they say, the infinite is different from them and is their source.

It is impossible, however, that there should be such a body; not because it is infinite – on that point a general proof can be given which applies equally to all, air, water, or anything else – but simply because there is, as a matter of fact, no such *sensible* body, alongside the so-called elements. Everything can be resolved into the elements of which it is composed. Hence the body in question would have been present in our world here, alongside air and fire and earth and water: but nothing of the kind is observed.

(*b*) Nor can fire or any other of the elements be infinite. For generally, and apart from the question how any of them could be infinite, the All, even if it were limited, cannot either be or become one of them, as Heraclitus says that at some time all things become fire. (The same argument applies also to the one which the physicists suppose to exist alongside the elements: for everything changes from contrary to contrary, e.g. from hot to cold).

The preceding consideration of the various cases serves to show us whether it is or is not possible that there should be an infinite sensible body. The following arguments give a general demonstration that it is not possible.

It is the nature of every kind of sensible body to be somewhere, and there is a place appropriate to each, the same for the part and for the whole, e.g. for the whole earth and for a single clod, and for fire and for a spark.

Suppose (*a*) that the infinite sensible body is homogeneous. Then each part will be either immovable or always being carried along. Yet neither is possible. For why downwards rather than upwards or in any other direction? I mean, e.g., if you take a clod, where will it be moved or where will it be at rest? For *ex hypothesi* the place of the body akin to it is infinite. Will it occupy the whole place, then? And how? What then will be the nature of its rest and of its movement, or where will they be? It will either be at home everywhere – then it will not be moved; or it will be moved everywhere – then it will not come to rest.

But if (*b*) the All has dissimilar parts, the proper places of the parts will be dissimilar also, and the body of the All will have no unity except that of contact. Then, further, the parts will be either finite or infinite in variety of kind. (i) *Finite* they cannot be, for if the All is to be infinite, some of them would have to be infinite, while the others were not, e.g. fire or water will be infinite. But, as we have seen before, such an element would destroy what is contrary to it. (This indeed is the reason why none of the physicists made fire or earth the one infinite body, but either water or air or what is intermediate between them, because the abode of each of the two was plainly determinate, while the others have an ambiguous place between up and down.)

But (ii) if the parts are *infinite* in number and simple, their proper places too will be infinite in number, and the same will be true of the elements themselves. If that is impossible, and the places are finite, the whole too must be finite; for the place and the body cannot but fit each other. Neither is the whole place larger than what can be filled by the body' (and then the body would no longer be infinite), nor is the body larger than the place; for either there would be an empty space or a body whose nature it is to be nowhere.

Anaxagoras gives an absurd account of why the infinite is at rest. He says that the infinite itself is the cause of its being fixed. This because it is *in* itself, since nothing else contains it – on the assumption that wherever anything is, it is there by its own nature. But this is not true: a thing could be somewhere by compulsion, and not where it is its nature to be.

Even if it is true as true can be that the whole is not moved (for what is fixed by itself and is in itself must be immovable), yet we must explain *why* it is not its nature to be moved. It is not enough just to make this statement and then decamp. Anything else might be in a state of rest, but there is no reason why it should not be its nature to be moved. The earth is not carried along, and would not be carried along if it were infinite, provided it is held together by the centre. But it would not be because there was no other region in which it could be carried along that it would remain at the centre, but because this is its nature. Yet in this case also we may say that it fixes itself. If then in the case of the earth, supposed to be infinite, it is at rest, not because it is infinite, but because it has weight and what is heavy rests at the centre and the earth is at the centre, similarly the infinite also would rest in itself, not because it is infinite and fixes itself, but owing to some other cause.

Another difficulty emerges at the same time. Any part of the infinite body ought to remain at rest. Just as the infinite remains at rest in itself because it fixes itself, so too any part of it you may take will remain in itself. The appropriate places of the whole and of the part are alike, e.g. of the whole earth and of a clod the appropriate place is the lower region; of fire as a whole and of a spark, the upper region. If, therefore, to be in itself is the place of the infinite, that also will be appropriate to the part. Therefore it will remain in itself.

In general, the view that there is an infinite body is plainly incompatible with the doctrine that there is necessarily a proper place for each kind of body, if every sensible body has either weight or lightness, and if a body has a natural locomotion towards the centre if it is heavy, and upwards if it is light. This would need to be true of the infinite also. But neither character can belong to it: it cannot be either as a whole, nor can it be half the one and half the other. For how should you divide it? Or how can the infinite have the one part up and the other down, or an extremity and a centre?

Further, every sensible body is in place, and the kinds or differences of place are up-down, before-behind, right-left; and these distinctions hold not only in relation to us and by arbitrary agreement, but also in the whole itself. But in the infinite body they cannot exist. In general, if it is impossible that there should be an infinite place, and if every body is in place, there cannot be an infinite body.

Surely what is in a special place is in place, and what is in place is in a special place. Just, then, as the infinite cannot be quantity – that would imply that it has a particular quantity, e.g. two or three cubits; quantity just means these – so a thing's being in place means that it is *somewhere*, and that is either up or down or in some other of the six differences of position: but each of these is a limit.

It is plain from these arguments that there is no body which is *actually* infinite.

That the Infinite Exists and How It Exists

But on the other hand to suppose that the infinite does not exist in any way leads obviously to many impossible consequences: there will be a beginning and an end of time, a magnitude will not be divisible into magnitudes, number will not be infinite. If, then, in view of the above considerations, neither alternative seems possible, an arbiter must be called in; and clearly there is a sense in which the infinite exists and another in which it does not.

We must keep in mind that the word 'is' means either what *potentially* is or what *fully* is.

Further, a thing is infinite either by addition or by division.

Now, as we have seen, magnitude is not actually infinite. But by division it is infinite. (There is no difficulty in refuting the theory of indivisible lines.) The alternative then remains that the infinite has a potential existence.

But the phrase 'potential existence' is ambiguous. When we speak of the potential existence of a statue we mean that there will be an actual statue. It is not so with the infinite. There will not be an actual infinite. The word 'is' has many senses, and we say that the infinite 'is' in the sense in which we say 'it is day' or 'it is the games', because one thing after another is always coming into existence. For of these things too the distinction between potential and actual existence holds. We say that there are Olympic games, both in the sense that they may occur and that they are actually occurring.

The infinite exhibits itself in different ways – in time, in the generations of man, and in the division of magnitudes. For generally the infinite has this mode of existence: one thing is always being taken after another, and each thing that is taken is always finite, but always different. Again, 'being' has more than one sense, so that we must not regard the infinite as a 'this', such as a man or a horse, but must suppose it to exist in the sense in which we speak of the day or the games as existing – things whose being has not come to them like that of a substance, but consists in a process of coming to be or passing away; definite if you like at each stage, yet always different.

But when this takes place in spatial magnitudes, what is taken persists, while in the succession of time and of men it takes place by the passing away of these in such a way that the source of supply never gives out.

In a way the infinite by addition is the same thing as the infinite by division. In a finite magnitude, the infinite by addition comes about in a way inverse to that of the other. For in proportion as we see division going on, in the same proportion we see addition being made to what is already marked off. For if we take a determinate part of a finite magnitude and add another part *determined by the same ratio* (not taking in the same amount of the original whole), and so on, we shall not traverse the given magnitude. But if we increase the ratio of the part, so as always to take in the same amount, we shall traverse the magnitude, for every finite magnitude is exhausted by means of any determinate quantity however small.

The infinite, then, exists in no other way, but in this way it does exist, potentially and by reduction. It exists fully in the sense in which we say 'it is day' or 'it is the games'; and potentially as matter exists, not independently as what is finite does.

By addition then, also, there is potentially an infinite, namely, what we have described as being in a sense the same as the infinite in respect of division. For it will always be possible to take something *ab extra*. Yet the sum of the parts taken will not exceed every determinate magnitude, just as in the direction of division every determinate magnitude is surpassed in smallness and there will be a smaller part.

But in respect of addition there cannot be an infinite which even potentially exceeds every assignable magnitude, unless it has the attribute of being actually infinite, as the physicists hold to be true of the body which is outside the world, whose essential nature is air or something of the kind. But if there cannot be in this way a sensible body which is infinite in the full sense, evidently there can no more be a body which is potentially infinite in respect of addition, except as the inverse of the infinite by division, as we have said. It is for this reason that Plato also made the infinites two in number, because it is supposed to be possible to exceed all limits and to proceed *ad infinitum* in the direction both of increase and of reduction. Yet though he makes the infinites two, he does not use them. For in the numbers the infinite in the direction of reduction is not

107

present, as the monad is the smallest; nor is the infinite in the direction of increase, for the parts number only up to the decad.

What the Infinite Is

The infinite turns out to be the contrary of what it is said to be. It is not what has nothing outside it that is infinite, but what always has something outside it. This is indicated by the fact that rings also that have no bezel are described as 'endless', because it is always possible to take a part which is outside a given part. The description depends on a certain similarity, but it is not true in the full sense of the word. This condition alone is not sufficient: it is necessary also that the next part which is taken should never be the same. In the circle, the latter condition is not satisfied: it is only the adjacent part from which the new part is different.

Our definition then is as follows:

A quantity is infinite if it is such that we can always take a part outside what has been already taken. On the other hand, what has nothing outside it is complete and whole. For thus we define the whole – that from which nothing is wanting, as a whole man or a whole box. What is true of each particular is true of the whole as such – the whole is that of which nothing is outside. On the other hand that from which something is absent and outside, however small that may be, is not 'all'. 'Whole' and 'complete' are either quite identical or closely akin. Nothing is complete (τέλειον) which has no end (τέλος) and the end is a limit.

Hence Parmenides must be thought to have spoken better than Melissus. The latter says that the whole is infinite, but the former describes it as limited, 'equally balanced from the middle'. For to connect the infinite with the all and the whole is not like joining two pieces of string; for it is from this they get the dignity they ascribe to the infinite – its containing all things and holding the all in itself – from its having a certain similarity to the whole. It is in fact the matter of the completeness which belongs to size, and what is potentially a whole, though not in the full sense. It is divisible both in the direction of reduction and of the inverse addition. It is a whole and limited; not, however, in virtue of its own nature, but in virtue of what is other than it. It does not contain, but, in so far as it is infinite, is contained. Consequently, also, it is unknowable, *qua* infinite; for the matter has no form. (Hence it is plain that the infinite stands in the relation of part rather than of whole. For the matter is part of the whole, as the bronze is of the bronze statue.) If it contains in the case of sensible things, in the case of intelligible things the great and the small ought to contain them. But it is absurd and impossible to suppose that the unknowable and indeterminate should contain and determine.

The Various Kinds of Infinite

It is reasonable that there should not be held to be an infinite in respect of addition such as to surpass every magnitude, but that there should be thought to be such an infinite in the direction of division. For the matter and the infinite are contained inside what contains them, while it is the form which contains. It is natural too to suppose that in number there is a limit in the direction of the minimum, and that in the other direction every assigned number is surpassed. In magnitude, on the contrary, every assigned magnitude is surpassed in the direction of smallness, while in the other direction there is no infinite magnitude. The reason is that what is one is indivisible whatever it may be, e.g. a man is one man, not many. Number on the other hand is a plurality of 'ones' and a certain quantity of them. Hence number must stop at the indivisible: for 'two' and 'three' are merely derivative terms, and so with each of the other numbers. But in the direction of largeness it is always possible to think of a larger number: for the number of times a magnitude can be bisected is infinite. Hence this infinite is potential, never actual: the number of parts that can be taken always surpasses any assigned number. But this number is not separable from the process of bisection, and its infinity is not a permanent actuality but consists in a process of coming to be, like time and the number of time.

With magnitudes the contrary holds. What is continuous is divided *ad infinitum*, but there is no infinite in the direction of increase. For the size which it can potentially be, it can also actually be. Hence since no sensible magnitude is infinite, it is impossible to exceed every assigned magnitude; for if it were possible there would be something bigger than the heavens.

The infinite is not the same in magnitude and movement and time, in the sense of a single nature, but its secondary sense depends on its primary sense, i.e. movement is called infinite in virtue of the magnitude covered by the movement (or alteration or growth), and time because of the movement. (I use these terms for the moment. Later I shall explain what each of them means, and also why every magnitude is divisible into magnitudes.)

Our account does not rob the mathematicians of their science, by disproving the actual existence of the infinite in the direction of increase, in the sense of the untraversable. In point of fact they do not need the infinite and do not use it. They postulate only that the finite straight line may be produced as far as they wish. It is possible to have divided in the same ratio as the largest quantity another magnitude of any size you like. Hence, for the purposes of proof, it will make no difference to them to have such an infinite instead, while its existence will be in the sphere of real magnitudes.

Which of the Four Conditions of Change the Infinite Is to Be Referred To

In the four-fold scheme of causes, it is plain that the infinite is a cause in the sense of matter, and that its essence is privation, the subject as such being what is continuous and sensible. All the other thinkers, too, evidently treat the infinite as matter – that is why it is inconsistent in them to make it what contains, and not what is contained.

Refutation of the Arguments for an Actual Infinite

It remains to dispose of the arguments which are supposed to support the view that the infinite exists not only potentially but as a separate thing. Some have no cogency; others can be met by fresh objections that are valid.

111

(1) In order that coming to be should not fail, it is not necessary that there should be a sensible body which is actually infinite. The passing away of one thing may be the coming to be of another, the All being limited.

(2) There is a difference between touching and being limited. The former is relative to something and is the touching of something (for everything that touches touches something), and further is an attribute of some one of the things which are limited. On the other hand, what is limited is not limited in relation to anything. Again, contact is not necessarily possible between any two things taken at random.

(3) To rely on mere thinking is absurd, for then the excess or defect is not in the thing but in the thought. One might think that one of us is bigger than he is and magnify him *ad infinitum*. But it does not follow that he is bigger than the size we are, just because someone thinks he is, but only because he *is* the size he is. The thought is an accident.

(*a*) Time indeed and movement are infinite, and also thinking, in the sense that each part that is taken passes in succession out of existence.

(*b*) Magnitude is not infinite either in the way of reduction or of magnification in thought.

This concludes my account of the way in which the infinite exists, and of the way in which it does not exist, and of what it is.

Book IV

Place

Does Place Exist?

The physicist must have a knowledge of Place, too, as well as of the infinite – namely, whether there is such a thing or not, and the manner of its existence and what it is – both because all suppose that things which exist are *somewhere* (the non-existent is nowhere – where is the goat-stag or the sphinx?), and because 'motion' in its most general and primary sense is change of place, which we call 'locomotion'.

The question 'what is place?' presents many difficulties. An examination of all the relevant facts seems to lead to divergent conclusions. Moreover, we have inherited nothing from previous thinkers, whether in the way of a statement of difficulties or of a solution.

The existence of place is held to be obvious from the fact of mutual replacement. Where water now is, there in turn, when the water has gone out as from a vessel, air is present. When therefore another body occupies this same place, the place is thought to be different from all the bodies which come to be in it and replace one another. What now contains air formerly contained water, so that clearly the place or space into which and out of which they passed was something different from both.

Further, the typical locomotions of the elementary natural bodies – namely, fire, earth, and the like – show not only that place is something, but also that it exerts a certain influence. Each is carried to its own place, if it is not hindered, the one up, the other down. Now these are regions or kinds of place – up and down and the rest of the six directions. Nor do such distinctions (up and down and right and left, &c.) hold only in relation to us. To us they are not always the same but change with the direction in which we are turned: that is why the same thing may be both right *and* left, up *and* down, before *and* behind. But in *nature* each is distinct, taken apart by itself. It is not every chance direction which is 'up', but where fire and what is light are carried; similarly, too, 'down' is not any chance direction but where what has weight and what is made of earth are carried – the implication being that these places do not differ merely in relative position, but also as possessing distinct potencies. This is made plain also by the objects studied by mathematics. Though they have no real place, they nevertheless, in respect of their position relatively to us, have a right and left as attributes ascribed to them only in consequence of their relative position, not having by nature these various characteristics. Again, the theory that the void exists involves the existence of place: for one would define void as place bereft of body.

These considerations then would lead us to suppose that place is something distinct from bodies, and that every sensible body is in place. Hesiod too might be held to have given a correct account of it when he made chaos first. At least he says:

First of all things came chaos to being, then broad-breasted earth, implying that things need to have space first, because he thought, with most people, that everything is somewhere and in place. If this is its nature, the potency of place must be a marvellous thing, and take precedence of all other things. For that without which nothing else can exist, while it can exist without the others, must needs be first; for place does not pass out of existence when the things in it are annihilated.

Doubts About the Nature of Place

True, but even if we suppose its existence settled, the question of its *nature* presents difficulty – whether it is some sort of 'bulk' of body or some entity other than that, for we must first determine its genus.

(1) Now it has three dimensions, length, breadth, depth, the dimensions by which all body also is bounded. But the place cannot *be* body; for if it were there would be two bodies in the same place.

(2) Further, if body has a place and space, clearly so too have surface and the other limits of body; for the same

statement will apply to them: where the bounding planes of the water were, there in turn will be those of the air. But when we come to a point we cannot make a distinction between it and its place. Hence if the place of a point is not different from the point, no more will that of any of the others be different, and place will not be something different from each of them.

(3) What in the world then are we to suppose place to be? If it has the sort of nature described, it cannot be an element or composed of elements, whether these be corporeal or incorporeal: for while it has size, it has not body. But the elements of sensible bodies are bodies, while nothing that has size results from a combination of intelligible elements.

(4) Also we may ask: of what in things is space the cause? None of the four modes of causation can be ascribed to it. It is neither cause in the sense of the matter of existents (for nothing is composed of it), nor as the form and definition of things, nor as end, nor does it move existents.

(5) Further, too, if it is itself an existent, *where* will it be? Zeno's difficulty demands an explanation: for if everything that exists has a place, place too will have a place, and so on *ad infinitum.*

(6) Again, just as every body is in place, so, too, every place has a body in it. What then shall we say about *growing* things? It follows from these premises that their place must grow with them, if their place is neither less nor greater than they are.

By asking these questions, then, we must raise the whole problem about place – not only as to what it is, but even whether there is such a thing.

Is Place Matter or Form?

We may distinguish generally between predicating B of A because it (A) is itself, and because it is something else; and particularly between place which is common and in which all bodies are, and the special place occupied primarily by each. I mean, for instance, that you are now in the heavens because you are in the air and it is in the heavens; and you are in the air because you are on the earth; and similarly on the earth because you are in this place which contains no more than you.

Now if place is what *primarily* contains each body, it would be a limit, so that the place would be the form or shape of each body by which the magnitude or the matter of the magnitude is defined: for this is the limit of each body.

If, then, we look at the question in this way the place of a thing is its form. But, if we regard the place as the *extension* of the magnitude, it is the matter. For this is different from the magnitude: it is what is contained and defined by the form, as by a bounding plane. Matter or the indeterminate is of this nature; when the boundary and attributes of a sphere are taken away, nothing but the matter is left.

This is why Plato in the *Timaeus* says that matter and space are the same; for the 'participant' and space are identical. (It is true, indeed, that the account he gives there of the 'participant' is different from what he says in his so-called 'unwritten teaching'. Nevertheless, he did identify place and space.) I mention Plato because, while all hold place to be something, he alone tried to say *what* it is.

117

In view of these facts we should naturally expect to find difficulty in determining what place is, if indeed it *is* one of these two things, matter or form. They demand a very close scrutiny, especially as it is not easy to recognize them apart.

But it is at any rate not difficult to see that place cannot be either of them. The form and the matter are not separate from the thing, whereas the place can be separated. As we pointed out, where air was, water in turn comes to be, the one replacing the other; and similarly with other bodies. Hence the place of a thing is neither a part nor a state of it, but is separable from it. For place is supposed to be something like a vessel – the vessel being a transportable place. But the vessel is no part of the thing.

In so far then as it is separable from the thing, it is not the form: *qua* containing, it is different from the matter.

Also it is held that what is anywhere is both itself something and that there is a different thing outside it. (Plato of course, if we may digress, ought to tell us why the form and the numbers are not in place, if 'what participates' is place – whether what participates is the Great and the Small or the matter, as he called it in writing in the *Timaeus*.)

Further, how could a body be carried to its own place, if place was the matter or the form? It is impossible that what has no reference to motion or the distinction of up and down can be place. So place must be looked for among things which have these characteristics.

If the place is in the thing (it must be if it is either shape or matter) place will have a place: for both the form and the indeterminate undergo change and motion along with the thing, and are not always in the same place, but are where the thing is. Hence the place will have a place.

Further, when water is produced from air, the place has been destroyed, for the resulting body is not in the same place. What sort of destruction then is that?

This concludes my statement of the reasons why space must be something, and again of the difficulties that may be raised about its essential nature.

Can a Thing Be in Itself or a Place Be in a Place?

The next step we must take is to see in how many senses one thing is said to be 'in' another.

(1) As the finger is 'in' the hand and generally the part is 'in' the whole.

(2) As the whole is 'in' the parts: for there is no whole over and above the parts.

(3) As man is 'in' animal and generally species 'in' genus.

(4) As the genus is 'in' the species and generally the part of the specific form 'in' the definition of the specific form.

(5) As health is 'in' the hot and the cold and generally the form 'in' the matter.

(6) As the affairs of Greece centre 'in' the king, and generally events centre 'in' their primary motive agent.

(7) As the existence of a thing centres 'in' its good and generally 'in' its end, i.e. in 'that for the sake of which' it exists.

119

(8) In the strictest sense of all, as a thing is 'in' a vessel, and generally 'in' place.

One might raise the question whether a thing can be in itself, or whether nothing can be in itself – everything being either *no*where or in something *else*.

The question is ambiguous; we may mean the thing *qua* itself or *qua* something else.

When there are parts of a whole – the one that in which a thing is, the other the thing which is in it – the whole will be described as being in itself. For a thing is described in terms of its parts, as well as in terms of the thing as a whole, e.g. a man is said to be white because the visible surface of him is white, or to be scientific because his thinking faculty has been trained. The jar then will not be in itself and the wine will not be in itself. But the jar of wine will: for the contents and the container are both parts of the same whole.

In this sense then, but not primarily, a thing can be in itself, namely, as 'white' is in body (for the visible surface is in body), and science is in the mind.

It is from these, which are 'parts' (in the sense at least of being 'in' the man), that the man is called white, &c. But the jar and the wine in separation are not parts of a whole, though together they are. So when there are parts, a thing will be in itself, as 'white' is in man because it is in body, and in body because it resides in the visible surface. We cannot go further and say that it is in surface in virtue of something other than itself. (Yet it is not in itself: though these are in a way the same thing,) they differ in essence, each having a special nature and capacity, 'surface' and 'white'.

Thus if we look at the matter inductively we do not find anything to be 'in' itself in any of the senses that have been distinguished; and it can be seen by argument that it is impossible. For each of two things will have to be both, e.g. the jar will have to be both vessel and wine, and the wine both wine and jar, if it is possible for a thing to be in itself; so that, however true it might be that they were in each other, the jar will receive the wine in virtue not of *its* being wine but of the wine's being wine, and the wine will be in the jar in virtue not of *its* being a jar but of the jar's being a jar. Now that they are different in respect of their essence is evident; for 'that in which something is' and 'that which is in it' would be differently defined.

Nor is it possible for a thing to be in itself even incidentally: for two things would be at the same time in the same thing. The jar would be in itself – if a thing whose nature it is to receive can be in itself; and that which it receives, namely (if wine) wine, will be in it.

Obviously then a thing cannot be in itself *primarily*.

Zeno's problem – that if Place is something it must be in something – is not difficult to solve. There is nothing to prevent the first place from being 'in' something else – is not indeed in that as 'in' place, but as health is 'in' the hot as a positive determination of it or as the hot is 'in' body as an affection. So we escape the infinite regress.

Another thing is plain: since the vessel is no part of what is in it (what contains in the *strict* sense is different from what is contained), place could not be either the matter or the form of the thing contained, but must be different – for the latter, both the matter and the shape, are parts of what is contained.

This then may serve as a critical statement of the difficulties involved.

What Place Is

What then after all is place? The answer to this question may be elucidated as follows.

Let us take for granted about it the various characteristics which are supposed correctly to belong to it essentially. We assume then –

(1) Place is what contains that of which it is the place.

(2) Place is no part of the thing.

(3) The immediate place of a thing is neither less nor greater than the thing.

(4) Place can be left behind by the thing and is separable. In addition:

(5) All place admits of the distinction of up and down, and each of the bodies is naturally carried to its appropriate place and rests there, and this makes the place either up or down.

Having laid these foundations, we must complete the theory. We ought to try to make our investigation such as will render an account of place, and will not only solve the difficulties connected with it, but will also show that the attributes supposed to belong to it do really belong to it, and further will make clear the cause of the trouble and of the difficulties about it. Such is the most satisfactory kind of exposition.

First then we must understand that place would not have been thought of, if there had not been a special kind of motion, namely that with respect to place. It is chiefly for this reason that we suppose the heaven also to be in place, because it is in constant movement. Of this kind of change there are two species – locomotion on the one hand and, on the other, increase and diminution. For these too involve variation of place: what was then in this place has now in turn changed to what is larger or smaller.

Again, when we say a thing is 'moved', the predicate either (1) belongs to it actually, in virtue of its own nature, or (2) in virtue of something conjoined with it. In the latter case it may be either (*a*) something which by its own nature is capable of being moved, e.g. the parts of the body or the nail in the ship, or (*b*) something which is not in itself capable of being moved, but is *always* moved through its conjunction with something else, as 'whiteness' or 'science'. These have changed their place only because the subjects to which they belong do so.

We say that a thing is in the world, in the sense of in place, because it is in the air, and the air is in the world; and when we say it is in the air, we do not mean it is in every part of the air, but that it is in the air because of the outer surface of the air which surrounds it; for if all the air were its place, the place of a thing would not be equal to the thing – which it is supposed to be, and which the primary place in which a thing is actually is.

When what surrounds, then, is not separate from the thing, but is in continuity with it, the thing is said to be in what surrounds it, not in the sense of in place, but as a part in a whole. But when the thing is separate and in contact, it is immediately 'in' the inner surface of the surrounding body, and this surface is neither a part of what is in it nor yet greater than its extension, but equal to it; for the extremities of things which touch are coincident.

Further, if one body is in continuity with another, it is not moved *in* that but *with* that. On the other hand it is moved *in* that if it is separate. It makes no difference whether what contains is moved or not.

Again, when it is not separate it is described as a part in a whole, as the pupil in the eye or the hand in the body: when it is separate, as the water in the cask or the wine in the jar. For the hand is moved *with* the body and the water *in* the cask.

It will now be plain from these considerations what place is. There are just four things of which place must be one – the shape, or the matter, or some sort of extension between the bounding surfaces of the containing body, or this boundary itself if it contains no extension over and above the bulk of the body which comes to be in it.

Three of these it obviously cannot be:

(1) The shape is supposed to be place because it surrounds, for the extremities of what contains and of what is contained are coincident. Both the shape and the place, it is true, are boundaries. But not of the same thing: the form is the boundary of the thing, the place is the boundary of the body which contains it.

(2) The extension between the extremities is thought to be something, because what is contained and separate may often

124

be changed while the container remains the same (as water may be poured from a vessel) – the assumption being that the extension is something over and above the body displaced. But there is no such extension. One of the bodies which change places and are naturally capable of being in contact with the container falls in – whichever it may chance to be.

If there were an extension which were such as to exist independently and be permanent, there would be an infinity of places in the same thing. For when the water and the air change places, all the portions of the two together will play the same part in the whole which was previously played by all the water in the vessel; at the same time the place too will be undergoing change; so that there will be another place which is the place of the place, and many places will be coincident. There is not a different place of the part, in which it is moved, when the whole vessel changes its place: it is always the same: for it is in the (proximate) place where they are that the air and the water (or the parts of the water) succeed each other, not in that place in which they come to be, which is part of the place which is the place of the whole world.

(3) The matter, too, might seem to be place, at least if we consider it in what is at rest and is thus separate but in continuity. For just as in change of quality there is something which was formerly black and is now white, or formerly soft and now hard – this is just why we say that the matter exists – so place, because it presents a similar phenomenon, is thought to exist – only in the one case we say so because *what* was air is now water, in the other because *where* air formerly was there

is now water. But the matter, as we said before, is neither separable from the thing nor contains it, whereas place has both characteristics.

Well, then, if place is none of the three – neither the form nor the matter nor an extension which is always there, different from, and over and above, the extension of the thing which is displaced – place necessarily is the one of the four which is left, namely, the boundary of the containing body which it is in contact with the contained body. (By the contained body is meant what can be moved by way of locomotion.)

Place is thought to be something important and hard to grasp, both because the matter and the shape present themselves along with it, and because the displacement of the body that is moved takes place in a stationary container, for it seems possible that there should be an interval which is other than the bodies which are moved. The air, too, which is thought to be incorporeal, contributes something to the belief: it is not only the boundaries of the vessel which seem to be place, but also what is between them, regarded as empty. Just, in fact, as the vessel is transportable place, so place is a non-portable vessel. So when what is within a thing which is moved, is moved and changes its place, as a boat on a river, what contains plays the part of a vessel rather than that of place. Place on the other hand is rather what is motionless: so it is rather the whole river that is place, because as a whole it is motionless.

Hence we conclude that *the innermost motionless boundary of what contains is place.*

This explains why the middle of the heaven and the surface which faces us of the rotating system are held to be 'up' and 'down' in the strict and fullest sense for all men: for the one is always at rest, while the inner side of the rotating body remains always coincident with itself. Hence since the light is what is naturally carried up, and the heavy what is carried down, the boundary which contains in the direction of the middle of the universe, and the middle itself, are down, and that which contains in the direction of the outermost part of the universe, and the outermost part itself, are up.

For this reason, too, place is thought to be a kind of surface, and as it were a vessel, i.e. a container of the thing.

Further, place is coincident with the thing, for boundaries are coincident with the bounded.

Corollaries

If then a body has another body outside it and containing it, it is in place, and if not, not. That is why, even if there were to be water which had not a container, the parts of it, on the one hand, will be moved (for one part is contained in another), while, on the other hand, the whole will be moved in one sense, but not in another. For as a whole it does not simultaneously change its place, though it will be moved in a circle: for this place is the place of its parts. (Some things are moved, not up and down, but in a circle; others up and down, such things namely as admit of condensation and rarefaction.)

As was explained, some things are potentially in place, others actually. So, when you have a homogeneous substance

127

which is continuous, the parts are potentially in place: when the parts are separated, but in contact, like a heap, they are actually in place.

Again, (1) some things are *per se* in place, namely every body which is movable either by way of locomotion or by way of increase is *per se* somewhere, but the heaven, as has been said, is not anywhere as a whole, nor in any place, if at least, as we must suppose, no body contains it. On the line on which it is moved, its parts have place: for each is contiguous to the next.

But (2) other things are in place indirectly, through something conjoined with them, as the soul and the heaven. The latter is, in a way, in place, for all its parts are: for on the orb one part contains another, That is why the upper part is moved in a circle, while the All is not anywhere. For what is somewhere is itself something, and there must be alongside it some other thing wherein it is and which contains it. But alongside the All or the Whole there is nothing outside the All, and for this reason all things are in the heaven; for the heaven, we may say, is the All. Yet their place is not the same as the heaven. It is part of it, the innermost part of it, which is in contact with the movable body; and for this reason the earth is in water, and this in the air, and the air in the aether, and the aether in heaven, but we cannot go on and say that the heaven is in anything else.

It is clear, too, from these considerations that all the problems which were raised about place will be solved when it is explained in this way:

(1) There is no necessity that the place should grow with the body in it,

(2) Nor that a point should have a place,

(3) Nor that two bodies should be in the same place,

(4) Nor that place should be a corporeal interval: for what is between the boundaries of the place is any body which may chance to be there, not an interval in body.

Further, (5) place is also somewhere, not in the sense of being in a place, but as the limit is in the limited; for not everything that is is in place, but only movable body.

Also (6) it is reasonable that each kind of body should be carried to its own place. For a body which is next in the series and in contact (not by compulsion) is akin, and bodies which are united do not affect each other, while those which are in contact interact on each other.

Nor (7) is it without reason that each should remain naturally in its proper place. For this part has the same relation to its place, as a separable part to its whole, as when one moves a part of water or air: so, too, air is related to water, for the one is like matter, the other form – water is the matter of air, air as it were the actuality of water, for water is potentially air, while air is potentially water, though in another way.

These distinctions will be drawn more carefully later. On the present occasion it was necessary to refer to them: what has now been stated obscurely will then be made more clear. If the matter and the fulfilment are the same thing (for water is both, the one potentially, the other completely), water will be related to air in a way as part to whole. That is why these have *contact*: it is *organic union* when both become actually one.

This concludes my account of place – both of its existence and of its nature.

The Void

The Views of Others About the Void

The investigation of similar questions about the void, also, must be held to belong to the physicist – namely whether it exists or not, and how it exists or what it is – just as about place. The views taken of it involve arguments both for and against, in much the same sort of way. For those who hold that the void exists regard it as a sort of place or vessel which is supposed to be 'full' when it holds the bulk which it is capable of containing, 'void' when it is deprived of that – as if 'void' and 'full' and 'place' denoted the same thing, though the essence of the three is different.

We must begin the inquiry by putting down the account given by those who say that it exists, then the account of those who say that it does not exist, and third the current view on these questions.

Those who try to show that the void does not exist do not disprove what people really mean by it, but only their erroneous way of speaking; this is true of Anaxagoras and of those who refute the existence of the void in this way. They merely give an ingenious demonstration that air is something – by straining wine-skins and showing the resistance of the air, and by cutting it off in clepsydras. But people really mean that there is an empty interval in which there is *no* sensible body. They hold that everything which is is body and say that what has nothing in it at all is void (so what is full of air is void). It is not then the existence of air that needs to be proved, but the non-existence of an interval, different from the bodies, either separable or actual – an interval which divides the whole body so as to break its continuity, as Democritus and Leucippus hold, and many other physicists – or even perhaps as something which is outside the whole body, which remains continuous.

These people, then, have not reached even the threshold of the problem, but rather those who say that the void exists.

(1) They argue, for one thing, that change in place (i.e. locomotion and increase) would not be. For it is maintained that motion would seem not to exist, if there were no void, since what is full cannot contain anything more. If it could, and there were two bodies in the same place, it would also be true that any number of bodies could be together; for it is impossible to draw a line of division beyond which the statement would become untrue. If this were possible, it would follow also that the smallest body would contain the greatest; for 'many a little makes a mickle': thus if many equal bodies can be together, so also can many unequal bodies.

Melissus, indeed, infers from these considerations that the All is immovable; for if it were moved there must, he says, be void, but void is not among the things that exist.

This argument, then, is one way in which they show that there is a void.

(2) They reason from the fact that some things are observed to contract and be compressed, as people say that a cask will hold the wine which formerly filled it, along with the skins into which the wine has been decanted, which implies that the compressed body contracts into the voids present in it.

Again (3) increase, too, is thought to take place always by means of void, for nutriment is body, and it is impossible for two bodies to be together. A proof of this they find also in what happens to ashes, which absorb as much water as the empty vessel.

The Pythagoreans too, (4) held that void exists and that it enters the heaven itself, which as it were inhales it, from the infinite air. Further it is the void which distinguishes the natures of things, as if it were like what separates and distinguishes the terms of a series. This holds primarily in the numbers, for the void distinguishes their nature.

These, then, and so many, are the main grounds on which people have argued for and against the existence of the void.

What 'Void' Means

As a step towards settling which view is true, we must determine the meaning of the name.

The void is thought to be place with nothing in it. The reason for this is that people take what exists to be body, and hold that while every body is in place, void is place in which there is no body, so that where there is no body, there must be void.

Every body, again, they suppose to be tangible; and of this nature is whatever has weight or lightness.

132

Hence, by a syllogism, what has nothing heavy or light in it, is void.

This result, then, as I have said, is reached by syllogism. It would be absurd to suppose that the point is void; for the void must be *place* which has in it an interval in tangible body.

But at all events we observe then that in one way the void is described as what is not full of body perceptible to touch; and what has heaviness and lightness is perceptible to touch. So we would raise the question: what would they say of an interval that has colour or sound – is it void or not? Clearly they would reply that if it *could* receive what is tangible it was void, and if not, not.

In another way void is that in which there is no 'this' or corporeal substance. So some say that the void is the matter of the body (they identify the place, too, with this), and in this they speak incorrectly; for the matter is not separable from the things, but they are inquiring about the void as about something separable.

Refutation of the Arguments for Belief in the Void

Since we have determined the nature of place, and void must, if it exists, be place deprived of body, and we have stated both in what sense place exists and in what sense it does not, it is plain that on this showing void does not exist, either unseparated or separated; for the void is meant to be, not body but rather an interval in body. This is why the void is thought to be something, viz. because place is, and for the same reasons. For the fact of motion in respect of place comes to the aid both of those who maintain that place is something over and above the bodies that come to occupy it, and of those who maintain that the void is something. They state that the void is the condition of movement in the sense of that in which movement takes place; and this would be the kind of thing that some say place is.

But there is no necessity for there being a void if there is movement. It is not in the least needed as a condition of movement in general, for a reason which, incidentally, escaped Melissus; viz. that the full can suffer *qualitative* change.

But not even movement in respect of place involves a void; for bodies may simultaneously make room for one another, though there is no interval separate and apart from the bodies that are in movement. And this is plain even in the rotation of continuous things, as in that of liquids.

And things can also be compressed not into a void but because they squeeze out what is contained in them (as, for instance, when water is compressed the air within it is squeezed out); and things can increase in size not only by the entrance of something but also by qualitative change; e.g. if water were to be transformed into air.

In general, both the argument about increase of size and that about the water poured on to the ashes get in their own way. For either not any and every part of the body is increased, or bodies may be increased otherwise than by the addition of body, or there may be two bodies in the same place (in which case they are claiming to solve a quite general difficulty, but are not proving the existence of void), or the *whole* body must be void, if it is increased in every part and is increased by means of void. The same argument applies to the ashes.

It is evident, then, that it is easy to refute the arguments by which they prove the existence of the void.

There Is No Void Separate from Bodies

Let us explain again that there is no void existing separately, as some maintain. If each of the simple bodies has a natural locomotion, e.g. fire upward and earth downward and towards the middle of the universe, it is clear that it cannot be the void that is the condition of locomotion. What, then, *will* the void be the condition of? It is thought to be the condition of movement in respect of place, and it is not the condition of this.

Again, if void is a sort of place deprived of body, when there is a void where will a body placed in it move to? It certainly cannot move into the whole of the void. The same argument applies as against those who think that place is something separate, into which things are carried; viz. how will what is placed in it move, or rest? Much the same argument will apply to the void as to the 'up' and 'down' in place, as is natural enough since those who maintain the existence of the void make it a place.

And in what way will things be present either in place or in the void? For the expected result does not take place when a body is placed as a whole in a place conceived of as separate and permanent; for a part of it, unless it be placed apart, will not be in a place but in the whole. Further, if separate place does not exist, neither will void.

If people say that the void must exist, as being necessary if there is to be movement, what rather turns out to be the case, if one studies the matter, is the opposite, that not a single thing can be moved if there *is* a void; for as with those who for a like reason say the earth is at rest, so, too, in the void things must be at rest; for there is no place to which things can move more or less than to another; since the void in so far as it is void admits no difference.

The second reason is this: all movement is either compulsory or according to nature, and if there is compulsory movement there must also be natural (for compulsory movement is contrary to nature, and movement contrary to nature is posterior to that according to nature, so that if each of the natural bodies has not a natural movement, none of the other movements can exist); but how can there be natural movement if there is no difference throughout the void or the infinite? For in so far as it is infinite, there will be no up or down or middle, and in so far as it is a void, up differs no whit from down; for as there is no difference in what is nothing, there is none in the void (for the void seems to be a non-existent and a privation of being), but natural locomotion seems to be differentiated, so that the things that exist by nature must be differentiated. Either, then, nothing has a natural locomotion, or else there is no void.

Further, in point of fact things that are thrown move though that which gave them their impulse is not touching them, either by reason of mutual replacement, as some maintain, or because the air that has been pushed pushes them with a movement quicker than the natural locomotion of the projectile wherewith it moves to its proper place. But in a void none of these things can take place, nor can anything be moved save as that which is carried is moved.

Further, no one could say why a thing once set in motion should stop anywhere; for why should it stop *here* rather than *here*? So that a thing will either be at rest or must be moved *ad infinitum*, unless something more powerful get in its way.

Further, things are now thought to move into the void because it yields; but in a void this quality is present equally everywhere, so that things should move in all directions.

Further, the truth of what we assert is plain from the following considerations. We see the same weight or body moving faster than another for two reasons, either because there is a difference in what it moves through, as between water, air, and earth, or because, other things being equal, the moving body differs from the other owing to excess of weight or of lightness.

Now the medium causes a difference because it impedes the moving thing, most of all if it is moving in the opposite direction, but in a secondary degree even if it is at rest; and especially a medium that is not easily divided, i.e. a medium that is somewhat dense.

A, then, will move through B in time Γ and through Δ, which is thinner, in time E (if the length of B is equal to Δ), in proportion to the density of the hindering body. For let B be water and Δ air; then by so much as air is thinner and more incorporeal than water, A will move through Δ faster than through B. Let the speed have the same ratio to the speed, then, that air has to water. Then if air is twice as thin, the body will traverse B in twice the time that it does Δ, and the time Γ will be twice the time E. And always, by so much as the medium is more incorporeal and less resistant and more easily divided, the faster will be the movement.

Now there is no ratio in which the void is exceeded by body, as there is no ratio of 0 to a number. For if 4 exceeds 3 by 1, and 2 by more than 1, and 1 by still more than it exceeds 2, still there is no ratio by which it exceeds 0; for that which exceeds must be divisible into the excess + that which is exceeded, so that 4 will be what it exceeds 0 by + 0. For this reason, too, a line does not exceed a point – unless it is composed of points! Similarly the void can bear no ratio to the full, and therefore neither can movement through the one to movement through the other, but if a thing moves through the thickest medium such and such a distance in such and such a time, it moves through the void with a speed beyond any ratio. For let Z be void, equal in magnitude to Band to Δ. Then if A is to traverse and move through it in a certain time, H, a time less than E, however, the void will bear this ratio to the full. But in a time equal to H, A will traverse the part Θ of Δ. And it will surely also traverse in that time any substance Z which exceeds air in thickness in the ratio which the time E bears to the time H. For if the body Z be as much thinner than Δ as E exceeds H, A, if it moves through Z, will traverse it in a time inverse to the speed of the movement, i.e. in a time equal to H. If, then, there is no body in Z, A will traverse Z still more quickly. But we supposed that its traverse of Z when Z was void occupied the time H. So that it will traverse Z in an equal time whether Z be full or void. But this is impossible. It is plain, then, that if there is a time in which it will move through any part of the void, this impossible result will follow: it will be found to traverse a certain distance, whether this be full or void, in an equal time; for there will be some *body* which is in the same ratio to the other body as the time is to the time.

To sum the matter up, the cause of this result is obvious, viz. that between any two movements there is a ratio (for they occupy time, and there is a ratio between any two times, so long as both are finite), but there is no ratio of void to full.

These are the consequences that result from a difference in the media; the following depend upon an excess of one moving body over another. We see that bodies which have a greater impulse either of weight or of lightness, if they are alike in other respects, move faster over an equal space, and in the ratio which their magnitudes bear to each other. Therefore they will also move through the void with this ratio of speed. But that is impossible; for why should one move faster? (In moving through *plena* it must be so; for the greater divides them faster by its force. For a moving thing cleaves the medium either by its shape, or by the impulse which the body that is carried along or is projected possesses.) Therefore all will possess equal velocity. But this is impossible.

It is evident from what has been said, then, that, if there is a void, a result follows which is the very opposite of the reason for which those who believe in a void set it up. They think that if movement in respect of place is to exist, the void cannot exist, separated all by itself; but this is the same as to say that place is a separate cavity; and this has already been stated to be impossible.

There Is No Void Occupied by Any Body

But even if we consider it on its own merits the so-called vacuum will be found to be really vacuous. For as, if one puts a cube in water, an amount of water equal to the cube will be displaced; so too in air; but the effect is imperceptible to sense. And indeed always, in the case of any body that can be displaced, it must, if it is not compressed, be displaced in the direction in which it is its nature to be displaced – always either down, if its locomotion is downwards as in the case of earth, or up, if it is fire, or in both directions – whatever be the nature of the inserted body. Now in the void this is impossible; for it is not body; the void must have penetrated the cube to a distance equal to that which this portion of void formerly occupied in the void, just as if the water or air had not been displaced by the wooden cube, but had penetrated right through it.

But the cube also has a magnitude equal to that occupied by the void; a magnitude which, if it is also hot or cold, or heavy or light, is none the less different in essence from all its attributes, even if it is not separable from them; I mean the *volume* of the wooden cube. So that even if it were separated from everything else and were neither heavy nor light, it will occupy an equal amount of void, and fill the same place, as the part of place or of the void equal to itself. How then will the body of the cube differ from the void or place that is equal to it? And if there can be two such things, why cannot there be any number coinciding?

This, then, is one absurd and impossible implication of the theory. It is also evident that the cube will have this same

140

volume even if it is displaced, which is an attribute possessed by all other bodies also. Therefore if this differs in no respect from its place, why need we assume a place for bodies over and above the volume of each, if their volume be conceived of as free from attributes? It contributes nothing to the situation if there is an equal interval attached to it as well. [Further, it ought to be clear by the study of moving things what sort of thing void is. But in fact it is found nowhere in the world. For air is something, though it does not *seem* to be so – nor, for that matter, would water, if fishes were made of iron; for the discrimination of the tangible is by touch.]

It is clear, then, from these considerations that there is no separate void.

There Is No Void in Bodies

There are some who think that the existence of rarity and density shows that there is a void. If rarity and density do not exist, they say, neither can things contract and be compressed. But if this were not to take place, either there would be no movement at all, or the universe would bulge, as Xuthus said, or air and water must always change into equal amounts (e.g. if air has been made out of a cupful of water, at the same time out of an equal amount of air a cupful of water must have been made), or void must necessarily exist; for compression and expansion cannot take place otherwise.

Now, if they mean by the rare that which has many voids existing separately, it is plain that if void cannot exist separate any more than a place can exist with an extension all to itself,

141

neither can the rare exist in this sense. But if they mean that there is void, not separately existent, but still present in the rare, this is less impossible, yet, first, the void turns out not to be a condition of *all* movement, but only of movement upwards (for the rare is light, which is the reason why they say fire is rare); second, the void turns out to be a condition of movement not as that in which it takes place, but in that the void carries things up as skins by being carried up themselves carry up what is continuous with them. Yet how can void have a local movement or a place? For thus that into which void moves is till then void of a void.

Again, how will they explain, in the case of what is heavy, its movement downwards? And it is plain that if the rarer and more void a thing is the quicker it will move upwards, if it were completely void it would move with a maximum speed! But perhaps even this is impossible, that it should move at all; the same reason which showed that in the void all things are incapable of moving shows that the void cannot move, viz., the fact that the speeds are incomparable.

Since we deny that a void exists, but for the rest the problem has been truly stated, that *either* there will be no movement, if there is not to be condensation and rarefaction, *or* the universe will bulge, *or* a transformation of water into air will always be balanced by an equal transformation of air into water (for it is clear that the air produced from water is bulkier than the water): it is necessary therefore, if compression does not exist, *either* that the next portion will be pushed outwards and make the outermost part bulge, *or* that somewhere else there must be an equal amount of water produced out of air, so that the entire bulk of the whole may be equal, *or* that nothing moves. For when anything is displaced this will always happen, unless it comes round in a circle; but locomotion is not always circular, but sometimes in a straight line.

These then are the reasons for which they might say that there is a void; *our* statement is based on the assumption that there is a single matter for contraries, hot and cold and the other natural contrarieties, and that what exists actually is produced from a potential existent, and that matter is not separable from the contraries but its being is different, and that a single matter may serve for colour and heat and cold.

The same matter also serves for both a large and a small body. This is evident; for when air is produced from water, the same matter has become something different, not by acquiring an addition to it, but has become actually what it was potentially, and, again, water is produced from air in the same way, the change being sometimes from smallness to greatness, and sometimes from greatness to smallness. Similarly, therefore, if air which is large in extent comes to have a smaller volume, or becomes greater from being smaller, it is the matter which is potentially both that comes to be each of the two.

For as the same matter becomes hot from being cold, and cold from being hot, because it was potentially both, so too from hot it can become more hot, though nothing in the matter has become hot that was not hot when the thing was less hot; just as, if the arc or curve of a greater circle becomes that of a smaller, whether it remains the same or becomes a different

143

curve, convexity has not come to exist in anything that was not convex but straight (for differences of degree do not depend on an intermission of the quality); nor can we get any portion of a flame, in which both heat and whiteness are not present. So too, then, is the earlier heat related to the later. So that the greatness and smallness, also, of the sensible volume are extended, not by the matter's acquiring anything new, but because the matter is potentially matter for both states; so that the same thing is dense and rare, and the two qualities have one matter.

The dense is heavy, and the rare is light. [Again, as the arc of a circle when contracted into a smaller space does not acquire a new part which is convex, but what was there has been contracted; and as any part of fire that one takes will be hot; so, too, it is all a question of contraction and expansion of the same matter.] There are two types in each case, both in the dense and in the rare; for both the heavy and the hard are thought to be dense, and contrariwise both the light and the soft are rare; and weight and hardness fail to coincide in the case of lead and iron.

From what has been said it is evident, then, that void does not exist either separate (either absolutely separate or as a separate element in the rare) or potentially, unless one is willing to call the condition of movement void, whatever it may be. At that rate the matter of the heavy and the light, *qua* matter of them, would be the void; for the dense and the rare are productive of locomotion in virtue of *this* contrariety, and in virtue of their hardness and softness productive of passivity and impassivity, i.e. not of locomotion but rather of qualitative change.

So much, then, for the discussion of the void, and of the sense in which it exists and the sense in which it does not exist.

Time

Doubts About the Existence of Time

Next for discussion after the subjects mentioned is Time.

The best plan will be to begin by working out the difficulties connected with it, making use of the current arguments. First, does it belong to the class of things that exist or to that of things that do not exist? Then secondly, what is its nature? To start, then: the following considerations would make one suspect that it either does not exist at all or barely, and in an obscure way. One part of it has been and is not, while the other is going to be and is not yet. Yet time – both infinite time and any time you like to take – is made up of these. One would naturally suppose that what is made up of things which do not exist could have no share in reality.

Further, if a divisible thing is to exist, it is necessary that, when it exists, all or some of its part must exist. But of time some parts have been, while others have to be, and no part of it *is*, though it is divisible. For what is 'now' is not a part: a part is a measure of the whole, which must be made up of parts. Time, on the other hand, is not held to be made up of 'nows'.

Again, the 'now' which seems to bound the past and the future – does it always remain one and the same or is it always other and other? It is hard to say.

(1) If it is always different and different, and if none of the *parts* in time which are other and other are simultaneous (unless the one contains and the other is contained, as the shorter time is by the longer), and if the 'now' which is not, but formerly was, must have ceased-to-be at some time, the '*nows*' too cannot be simultaneous with one another, but the prior 'now' must always have ceasedto-be. But the prior 'now' cannot have ceased-to-be in itself (since it then existed); yet it cannot have ceased-to-be in another 'now'. For we may lay it down that one 'now' cannot be next to another, any more than point to point. If then it did not cease-to-be in the next 'now' but in another, it would exist simultaneously with the innumerable 'nows' between the two – which is impossible.

Yes, but (2) neither is it possible for the 'now' to remain always the same. No determinate divisible thing has a single termination, whether it is continuously extended in one or in more than one dimension: but the 'now' is a termination, and it is possible to cut off a determinate time. Further, if coincidence in time (i.e. being neither prior nor posterior) means to be 'in one and the same "now"', then, if both what is before and what is after are in this same 'now', things which happened ten thousand years ago would be simultaneous with what has happened today, and nothing would be before or after anything else.

This may serve as a statement of the difficulties about the attributes of time.

Various Opinions About the Nature of Time

As to what time is or what is its nature, the traditional accounts give us as little light as the preliminary problems which we have worked through.

Some assert that it is (1) the movement of the whole, others that it is (2) the sphere itself.

(1) Yet part, too, of the revolution is a time, but it certainly is not a revolution: for what is taken is part of a revolution, not a revolution. Besides, if there were more heavens than one, the movement of any of them equally would be time, so that there would be many times at the same time.

(2) Those who said that time is the sphere of the whole thought so, no doubt, on the ground that all things are in time and all things are in the sphere of the whole. The view is too naive for it to be worthwhile to consider the impossibilities implied in it.

But as time is most usually supposed to be (3) motion and a kind of change, we must consider this view.

Now (a) the change or movement of each thing is only in the thing which changes or *where* the thing itself which moves or changes may chance to be. But time is present equally everywhere and with all things.

Again, (b) change is always faster or slower, whereas time is not: for 'fast' and 'slow' are defined by time – 'fast' is what moves much in a short time, 'slow' what moves little in a long time; but time is not defined by time, by being either a certain amount or a certain kind of it.

Clearly then it is not movement. (We need not distinguish at present between 'movement' and 'change'.)

What Time Is

But neither does time exist without change; for when the state of our own minds does not change at all, or we have not noticed its changing, we do not realize that time has elapsed, any more than those who are fabled to sleep among the heroes in Sardinia do when they are awakened; for they connect the earlier 'now' with the later and make them one, cutting out the interval because of their failure to notice it. So, just as, if the 'now' were not different but one and the same, there would not have been time, so too when its difference escapes our notice the interval does not seem to be time. If, then, the non-realization of the existence of time happens to us when we do not distinguish any change, but the soul seems to stay in one indivisible state, and when we perceive and distinguish we say time has elapsed, evidently time is not independent of movement and change. It is evident, then, that time is neither movement nor independent of movement.

We must take this as our starting-point and try to discover – since we wish to know what time is – what exactly it has to do with movement.

Now we perceive movement and time together: for even when it is dark and we are not being affected through the body, if any movement takes place in the mind we at once suppose that some time also has elapsed; and not only that but also, when some time is thought to have passed, some movement also along with it seems to have taken place. Hence time is either movement or something that belongs to movement. Since then it is not movement, it must be the other.

But what is moved is moved from something to something, and all magnitude is continuous. Therefore the movement goes with the magnitude. Because the magnitude is continuous, the movement too must be continuous, and if the movement, then the time; for the time that has passed is always thought to be in proportion to the movement.

The distinction of 'before' and 'after' holds primarily, then, in place; and there in virtue of relative position. Since then 'before' and 'after' hold in magnitude, they must hold also in movement, these corresponding to those. But also in time the distinction of 'before' and 'after' must hold, for time and movement always correspond with each other. The 'before' and 'after' in motion identical substratum with motion yet differs from it in definition, and is not identical with motion.

But we apprehend time only when we have marked motion, marking it by 'before' and 'after'; and it is only when we have perceived 'before' and 'after' in motion that we say that time has elapsed. Now we mark them by judging that A and B are different, and that some third thing is intermediate to them. When we think of the extremes as different from the middle and the mind pronounces that the 'nows' are two, one before and one after, it is then that we say that there is time, and this that we say is time. For what is bounded by the 'now' is thought to be time – we may assume this.

When, therefore, we perceive the 'now' as one, and neither as before and after in a motion nor as an identity but in relation to a 'before' and an 'after', no time is thought to have elapsed, because there has been no motion either. On the other hand, when we do perceive a 'before' and an 'after', then we say that there is time. For time is just this – number of motion in respect of 'before' and 'after'.

Hence time is not movement, but only movement in so far as it admits of enumeration. A proof of this: we discriminate the more or the less by number, but more or less movement by time. Time then is a kind of number. (Number, we must note, is used in two senses – both of what is counted or the countable and also of that with which we count. Time obviously is what is counted, not that with which we count: these are different kinds of thing.)

The 'Now'

Just as motion is a perpetual succession, so also is time. But every simultaneous time is self-identical; for the 'now' as a subject is an identity, but it accepts different attributes. The 'now' measures time, in so far as time involves the 'before and after'.

The 'now' in one sense is the same, in another it is not the same. In so far as it is in succession, it is different (which is just what its being now was supposed to mean), but its substratum is an identity: for motion, as was said, goes with magnitude, and time, as we maintain, with motion. Similarly, then, there corresponds to the point the body which is carried along, and by which we are aware of the motion and of the 'before and after' involved in it. This is an identical *substratum* (whether a point or a stone or something else of the kind), but it has different *attributes* – as the sophists assume that Coriscus' being in the Lyceum is a different thing from Coriscus' being in the market-place. And the body which is carried along is different, in so far as it is at one time here and at another there. But the 'now' corresponds to the body that is carried along, as time corresponds to the motion. For it is by means of the body that is carried along that we become aware of the 'before and after' in the motion, and if we regard these as countable we get the 'now'. Hence in these also the 'now' as substratum remains the same (for it is what is before and after in movement), but what is predicated of it is different; for it is in so far as the 'before and after' is numerable that we get the 'now'. This is what is most knowable: for, similarly, motion is known because of that which is moved, locomotion because of that which is carried. For what is carried is a real thing, the movement is not. Thus what is called 'now' in one sense is always the same; in another it is not the same: for this is true also of what is carried.

Clearly, too, if there were no time, there would be no 'now', and *vice versa*. Just as the moving body and its locomotion involve each other mutually, so too do the number of the moving body and the number of its locomotion. For the number of the locomotion is time, while the 'now' corresponds to the moving body, and is like the unit of number.

Time, then, also is both made continuous by the 'now' and divided at it. For here too there is a correspondence with the locomotion and the moving body. For the motion or locomotion is made one by the thing which is moved, because *it* is one – not because it is one in its own nature (for there might be pauses in the movement of such a thing) – but because it is one in definition: for this determines the movement as 'before' and 'after'. Here, too, there is a correspondence with the point; for the point also both connects and terminates the length – it is the beginning of one and the end of another. But when you take it in this way, using the one point as two, a pause is necessary, if the same point is to be the beginning and the end. The 'now' on the other hand, since the body carried is moving, is always different.

Hence time is not number in the sense in which there is 'number' of the same point because it is beginning and end, but rather as the extremities of a line form a number, and not as the parts of the line do so, both for the reason given (for we can use the middle point as two, so that on that analogy time might stand still), and further because obviously the 'now' is no *part* of time nor the section any part of the movement, any more than the points are parts of the line – for it is two *lines* that are *parts* of one line.

In so far then as the 'now' is a boundary, it is not time, but an attribute of it; in so far as it numbers, it is number; for boundaries belong only to that which they bound, but number (e.g. ten) is the number of these horses, and belongs also elsewhere.

It is clear, then, that time is 'number of movement in respect of the before and after', and is continuous since it is an attribute of what is continuous.

Various Attributes of Time

The smallest number, in the strict sense of the word 'number', is two. But of number as concrete, sometimes there is a minimum, sometimes not: e.g. of a 'line', the smallest in respect of *multiplicity* is two (or, if you like, one), but in respect of size there is no minimum; for every line is divided *ad infinitum*. Hence it is so with time. In respect of number the minimum is one (or two); in point of extent there is no minimum.

It is clear, too, that time is not described as fast or slow, but as many or few and as long or short. For as continuous it is long or short and as a number many or few, but it is not fast or slow – any more than any number with which we number is fast or slow.

Further, there is the same time everywhere at once, but not the same time before and after, for while the present change is one, the change which has happened and that which will happen are different. Time is not number with which we count, but the number of things which are counted, and this according as it occurs before or after is always different, for the 'nows' are different. And the number of a hundred horses and a hundred men is the same, but the things numbered are different – the horses from the men. Further, as a movement can be one and the same again and again, so too can time, e.g. a year or a spring or an autumn.

Not only do we measure the movement by the time, but also the time by the movement, because they define each other. The time marks the movement, since it is its number, and the movement the time. We describe the time as much or little, measuring it by the movement, just as we know the number by what is numbered, e.g. the number of the horses by one horse as the unit. For we know how many horses there are by the use of the number; and again by using the one horse as unit we know the number of the horses itself. So it is with the time and the movement; for we measure the movement by the time and vice versa. It is natural that this should happen; for the movement goes with the distance and the time with the movement, because they are quanta and continuous and divisible. The movement has these attributes because the distance is of this nature, and the time has them because of the movement. And we measure both the distance by the movement and the movement by the distance; for we say that the road is long, if the journey is long, and that this is long, if the road is long – the time, too, if the movement, and the movement, if the time.

The Things that Are in Time

Time is a measure of motion and of being moved, and it measures the motion by determining a motion which will measure exactly the whole motion, as the cubit does the length by determining an amount which will measure out the whole. Further 'to be in time' means, for movement, that both it and its essence are measured by time (for simultaneously it measures both the movement and its essence, and this is what being in time means for it, that its essence should be measured).

Clearly then 'to be in time' has the same meaning for other things also, namely, that their being should be measured by time. 'To be in time' is one of two things: (1) to exist when time exists, (2) as we say of some things that they are 'in number'.

The latter means either what is a part or mode of number – in general, something which belongs to number – or that things have a number.

Now, since time is number, the 'now' and the 'before' and the like are in time, just as 'unit' and 'odd' and 'even' are in number, i.e. in the sense that the one set belongs to number, the other to time. But things are in time as they are in number. If this is so, they are contained by time as things in place are contained by place.

Plainly, too, to be in time does not mean to coexist with time, any more than to be in motion or in place means to coexist with motion or place. For if 'to be in something' is to mean this, then all things will be in anything, and the heaven will be in a grain; for when the grain is, then also is the heaven. But this is a merely incidental conjunction, whereas the other is necessarily involved: that which is in time necessarily involves that there is time when *it* is, and that which is in motion that there is motion when *it* is.

Since what is 'in time' is so in the same sense as what is in number is so, a time greater than everything in time can be found. So it is necessary that all the things in time should be contained by time, just like other things also which are 'in anything', e.g. the things 'in place' by place.

A thing, then, will be affected by time, just as we are accustomed to say that time wastes things away, and that all things grow old through time, and that there is oblivion owing to the lapse of time, but we do not say the same of getting to know or of becoming young or fair. For time is by its nature

155

the cause rather of decay, since it is the number of change, and change removes what is.

Hence, plainly, things which are always are not, as such, in time, for they are not contained by time, nor is their being measured by time. A proof of this is that none of them is *affected* by time, which indicates that they are not in time.

Since time is the measure of motion, it will be the measure of rest too – indirectly. For all rest is in time. For it does not follow that what is in time is moved, though what is in motion is necessarily moved. For time is not motion, but 'number of motion': and what is at rest, also, can be in the number of motion. Not everything that is not in motion can be said to be 'at rest' – but only that which can be moved, though it actually is not moved, as was said above.

'To be in number' means that there is a number of the thing, and that its being is measured by the number in which it is. Hence if a thing is 'in time' it will be measured by time. But time will measure what is moved and what is at rest, the one *qua* moved, the other *qua* at rest; for it will measure their motion and rest respectively.

Hence what is moved will not be measurable by the time simply in so far as it has quantity, but in so far as its *motion* has quantity. Thus none of the things which are neither moved nor at rest are in time: for 'to be in time' is 'to be measured by time', while time is the measure of motion and rest.

Plainly, then, neither will everything that does not exist be in time, i.e. those non-existent things that cannot exist, as the diagonal cannot be commensurate with the side.

Generally, if time is directly the measure of motion and indirectly of other things, it is clear that a thing whose existence is measured by it will have its existence in rest or motion. Those things therefore which are subject to perishing and becoming – generally, those which at one time exist, at another do not – are necessarily in time: for there is a greater time which will extend both beyond their existence and beyond the time which measures their existence. Of things which do not exist but are contained by time some were, e.g. Homer once was, some will be, e.g. a future event; this depends on the direction in which time contains them; if on both, they have both modes of existence. As to such things as it does not contain in any way, they neither were nor are nor will be. These are those non-existents whose opposites always are, as the incommensurability of the diagonal always is – and this will not be in time. Nor will the commensurability, therefore; hence this eternally is not, because it is contrary to what eternally is. A thing whose contrary is not eternal can be and not be, and it is of such things that there is coming to be and passing away.

Definitions of Temporal Terms

The 'now' is the link of time, as has been said (for it connects past and future time), and it is a limit of time (for it is the beginning of the one and the end of the other). But this is not obvious as it is with the point, which is fixed. It divides potentially, and in so far as it is dividing the 'now' is always different, but in so far as it connects it is always the same, as it is with mathematical lines. For the intellect it is not always one

157

and the same point, since it is other and other when one divides the line; but in so far as it is one, it is the same in every respect.

So the 'now' also is in one way a potential dividing of time, in another the termination of both parts, and their unity. And the dividing and the uniting are the same thing and in the same reference, but in essence they are not the same.

So one kind of 'now' is described in this way: another is when the time is *near* this kind of 'now'. 'He will come now' because he will come today; 'he has come now' because he came today. But the things in the *Iliad* have not happened 'now', nor is the flood 'now' – not that the time from now to them is not continuous, but because they are not near.

'At some time' means a time determined in relation to the first of the two types of 'now', e.g. 'at some time' Troy was taken, and 'at some time' there will be a flood; for it must be determined with reference to the 'now'. There *will* thus be a determinate time from this 'now' to that, and there *was* such in reference to the past event. But if there be no time which is not 'sometime', every time will be determined.

Will time then fail? Surely not, if motion always exists. Is time then always different or does the same time recur? Clearly time is, in the same way as motion is. For if one and the same motion sometimes recurs, it will be one and the same time, and if not, not.

Since the 'now' is an end and a beginning of time, not of the same time however, but the end of that which is past and the beginning of that which is to come, it follows that, as the circle has its convexity and its concavity, in a sense, in the same thing, so time is always at a beginning and at an end. And for this reason it seems to be always different; for the 'now' is not the beginning and the end of the same thing; if it were, it would be at the same time and in the same respect two opposites. And time will not fail; for it is always at a beginning.

'Presently' or 'just' refers to the part of future time, which is near the indivisible present 'now' ('When do you walk?' 'Presently', because the time in which he is going to do so is near), and to the part of past time which is not far from the 'now' ('When do you walk?' 'I have just been walking'). But to say that Troy has just been taken – we do not say that, because it is too far from the 'now'. 'Lately', too, refers to the part of past time which is near the present 'now'. 'When did you go?' 'Lately', if the time is near the existing now. 'Long ago' refers to the distant past.

'Suddenly' refers to what has departed from its former condition in a time imperceptible because of its smallness; but it is the nature of *all* change to alter things from their former condition. In time all things come into being and pass away; for which reason some called it the wisest of all things, but the Pythagorean Paron called it the most stupid, because in it we also forget; and his was the truer view. It is clear then that it must be in itself, as we said before, the condition of destruction rather than of coming into being (for change, in itself, makes things depart from their former condition), and only incidentally of coming into being, and of being. A sufficient evidence of this is that nothing comes into being without itself moving somehow and acting, but a thing can be destroyed even if it does not move at all. And this is what, as a rule, we chiefly mean by a thing's being destroyed by time. Still, time does not work even this change; even this sort of change takes place *incidentally* in time.

We have stated, then, that time exists and what it is, and in how many senses we speak of the 'now', and what 'at some time', 'lately', 'presently' or 'just', 'long ago', and 'suddenly' mean.

Further Reflections About Time

These distinctions having been drawn, it is evident that every change and everything that moves is in time; for the distinction of faster and slower exists in reference to all change, since it is found in every instance. In the phrase 'moving faster' I refer to that which changes before another into the condition in question, when it moves over the same interval and with a regular movement; e.g. in the case of locomotion, if both things move along the circumference of a circle, or both along a straight line; and similarly in all other cases. But what is *before* is in time; for we say 'before' and 'after' with reference to the distance from the 'now', and the 'now' is the boundary of the past and the future; so that since 'nows' are in time, the before and the after will be in time too; for in that in which the 'now' is, the distance from the 'now' will also be. But 'before' is used contrariwise with reference to past and to future time; for in the past we call 'before' what is farther from the 'now', and 'after' what is nearer, but in the future we call the nearer 'before' and the farther 'after'. So that since the 'before' is in time, and every movement involves a 'before', evidently every change and every movement is in time.

It is also worth considering how time can be related to the soul; and why time is thought to be in everything, both in earth and in sea and in heaven. Is it because it is an attribute, or state, of movement (since it is the number of movement) and all these things are movable (for they are all in place), and time and movement are together, both in respect of potentiality and in respect of actuality?

Whether if soul did not exist time would exist or not, is a question that may fairly be asked; for if there cannot be someone to count there cannot be anything that can be counted, so that evidently there cannot be number; for number is either what has been, or what can be, counted. But if nothing but soul, or in soul reason, is qualified to count, there would not be time unless there were soul, but only that of which time is an attribute, i.e. if *movement* can exist without soul, and the before and after are attributes of movement, and time is these *qua* numerable.

One might also raise the question what sort of movement time is the number of. Must we not say 'of *any* kind'? For things both come into being in time and pass away, and grow, and are altered in time, and are moved locally; thus it is of each movement *qua* movement that time is the number. And so it is simply the number of continuous movement, not of any particular kind of it.

But other things as well may have been moved now, and there would be a number of each of the two movements. Is there another time, then, and will there be two equal times at once?

Surely not. For a time that is both equal and simultaneous is one and the same time, and even those that are not simultaneous are one in kind; for if there were dogs, and horses, and seven of each, it would be the same number. So, too, movements that have simultaneous limits have the same time, yet the one may in fact be fast and the other not, and one may be locomotion and the other alteration; still the time of the two changes is the same if their number also is equal and simultaneous; and for this reason, while the movements are different and separate, the time is everywhere the same, because the number of equal and simultaneous movements is everywhere one and the same.

Now there is such a thing as locomotion, and in locomotion there is included circular movement, and everything is measured by some one thing homogeneous with it, units by a unit, horses by a horse, and similarly times by some definite time, and, as we said, time is measured by motion as well as motion by time (this being so because by a motion definite in time the quantity both of the motion and of the time is measured): if, then, what is first is the measure of everything homogeneous with it, regular circular motion is above all else the measure, because the number of this is the best known. Now neither alteration nor increase nor coming into being can be regular, but locomotion can be. This also is why time is thought to be the movement of the sphere, viz. because the other movements are measured by this, and time by this movement.

This also explains the common saying that human affairs form a circle, and that there is a circle in all other things that have a natural movement and coming into being and passing away. This is because all other things are discriminated by time, and end and begin as though conforming to a cycle; for even time itself is thought to be a circle. And this opinion again is held because time is the measure of this kind of locomotion and is itself measured by such. So that to say that the things that come into being form a circle is to say that there is a circle of time; and this is to say that it is measured by the circular movement; for apart from the measure nothing else to be measured is observed; the whole is just a plurality of measures.

It is said rightly, too, that the number of the sheep and of the dogs is the same *number* if the two numbers are equal, but not the same *decad* or the same *ten*; just as the equilateral and the scalene are not the same *triangle*, yet they are the same *figure*, because they are both triangles. For things are called the same so-and-so if they do not differ by a differentia of that thing, but not if they do; e.g. triangle differs from triangle by a differentia of triangle, therefore they are different triangles; but they do not differ by a differentia of figure, but are in one and the same division of it. For a figure of one kind is a circle and a figure of another kind a triangle, and a triangle of one kind is equilateral and a triangle of another kind scalene. They are the same figure, then, and that, triangle, but not the same triangle. Therefore the number of two groups also is the same number (for their number does not differ by a differentia of number), but it is not the same decad; for the things of which it is asserted differ; one group are dogs, and the other horses.

We have now discussed time – both time itself and the matters appropriate to the consideration of it.

Book V

Classification of Movements and Changes

Everything which changes does so in one of three senses. It may change (1) *accidentally*, as for instance when we say that something musical walks, that which walks being something in which aptitude for music is an *accident*. Again (2) a thing is said without qualification to change because *something belonging to it* changes, i.e. in statements which refer to part of the thing in question: thus the body is restored to health because the eye or the chest, that is to say a *part* of the whole body, is restored to health. And above all there is (3) the case of a thing which is in motion neither accidentally nor in respect of something else belonging to it, but in virtue of being *itself* directly in motion. Here we have a thing which is *essentially* movable: and that which is so is a different thing according to the particular variety of motion: for instance it may be a thing capable of alteration: and within the sphere of alteration it is again a different thing according as it is capable of being restored to health or capable of being heated. And there are the same distinctions in the case of the mover: (1) one thing causes motion accidentally, (2) another partially (because something belonging to it causes motion), (3) another of itself directly, as, for instance, the physician heals, the hand strikes. We have, then, the following factors: (*a*) on the one hand that which directly causes motion, and (*b*) on the other hand that which is in motion: further, we have (*c*) that in which motion takes place, namely time, and (distinct from these three) (*d*) that from which and (*e*) that to which it proceeds: for every motion proceeds from something and to something, that which is directly in motion being distinct from that to which it is in motion and that from which it is in motion: for instance, we may take the three things 'wood', 'hot', and 'cold', of which the first is that which is in motion, the second is that to which the motion proceeds, and the third is that from which it proceeds. This being so, it is clear that the motion is in the wood, not in its form: for the motion is neither caused nor experienced by the form or the place or the quantity. So we are left with a mover, a moved, and a goal of motion. I do not include the starting-point of motion: for it is the goal rather than the starting-point of motion that gives its name to a particular process of change. Thus 'perishing' is change *to not-being*, though it is also true that that which perishes changes *from being*: and 'becoming' is change *to being*, though it is also change *from not-being*.

165

Now a definition of motion has been given above, from which it will be seen that every goal of motion, whether it be a form, an affection, or a place, is immovable, as, for instance, knowledge and heat. Here, however, a difficulty may be raised. Affections, it may be said, are motions, and whiteness is an affection: thus there may be change *to* a motion. To this we may reply that it is not whiteness but whitening that is a motion. Here also the same distinctions are to be observed: a goal of motion may be so accidentally, or partially and with reference to something other than itself, or directly and with no reference to anything else: for instance, a thing which is becoming white changes accidentally to an object of thought, the *colour* being only accidentally the object of thought; it changes to colour, because white is a part of colour, or to Europe, because Athens is a part of Europe; but it changes essentially to white colour. It is now clear in what sense a thing is in motion essentially, accidentally, or in respect of something other than itself, and in what sense the phrase 'itself directly' is used in the case both of the mover and of the moved: and it is also clear that the motion is not in the form but in that which is in motion, that is to say 'the movable in activity'. Now accidental change we may leave out of account: for it is to be found in everything, at any time, and in any respect. Change which is not accidental on the other hand is not to be found in everything, but only in contraries, in things intermediate between contraries, and in contradictories, as may be proved by induction. An intermediate may be a starting-point of change, since for the purposes of the change it serves as contrary to either of two contraries: for the intermediate is in a sense the extremes. Hence we speak of the intermediate as in a sense a contrary relatively to the extremes and of either extreme as a contrary relatively to the intermediate: for instance, the central note is low relatively to the highest and high relatively to the lowest, and grey is light relatively to black and dark relatively to white.

Classification of Changes *per Se*

And since every change is *from* something *to* something – as the word itself (μεταβολή) indicates, implying something 'after' (μετά) something else, that is to say something earlier and something later – that which changes must change in one of four ways: from subject to subject, from subject to non-subject, from non-subject to subject, or from non-subject to non-subject, where by 'subject' I mean what is affirmatively expressed. So it follows necessarily from what has been said above that there are only three kinds of change, that from subject to subject, that from subject to non-subject, and that from non-subject to subject: for the fourth conceivable kind, that from non-subject to non-subject, is not change, as in that case there is no opposition either of contraries or of contradictories.

Now change from non-subject to subject, the relation being that of contradiction, is 'coming to be' – 'unqualified coming to be' when the change takes place in an unqualified way, 'particular coming to be' when the change is change in a particular character: for instance, a change from not-white to white is a coming to be of the particular thing, white, while change from unqualified not-being to being is coming to be in an unqualified way, in respect of which we say that a thing 'comes to be' without qualification, not that it 'comes to be' some particular thing. Change from subject to non-subject is 'perishing' – 'unqualified perishing' when the change is from being to not-being, 'particular perishing' when the change is to the opposite negation, the distinction being the same as that made in the case of coming to be.

Now the expression 'not-being' is used in several senses: and there can be motion neither of that which 'is not' in respect of the affirmation or negation of a predicate, nor of that which 'is not' in the sense that it only *potentially* 'is', that is to say the opposite of that which *actually* 'is' in an unqualified sense: for although that which is 'not-white ' or 'not-good' may nevertheless be in motion *accidentally* (for example that which is 'not-white' might be a man), yet that which is without qualification 'not-so-and-so' cannot in any sense be in motion: therefore it is impossible for that which *is not* to be in motion. This being so, it follows that 'becoming' cannot be a motion: for it is that which 'is not' that 'becomes'. For however true it may be that it *accidentally* 'becomes', it is nevertheless correct to say that it is that which 'is not' that in an unqualified sense 'becomes'. And similarly it is impossible for that which 'is not' to be at rest.

There are these difficulties, then, in the way of the assumption that that which 'is not' can be in motion: and it may be further objected that, whereas everything which is in motion is in space, that which 'is not' is not in space: for then it would be *somewhere*.

168

So, too, 'perishing' is not a motion: for a motion has for its contrary either another motion or rest, whereas 'perishing' is the contrary of 'becoming'.

Since, then, every motion is a kind of change, and there are only the three kinds of change mentioned above; and since of these three those which take the form of 'becoming' and 'perishing', that is to say those which imply a relation of contradiction, are not motions: it necessarily follows that only change from subject to subject is motion. And every such subject is either a contrary or an intermediate (for a privation may be allowed to rank as a contrary) and can be affirmatively expressed, as naked, toothless, or black. If, then, the categories are severally distinguished as Being, Quality, Place, Time, Relation, Quantity, and Activity or Passivity, it necessarily follows that there are three kinds of motion – qualitative, quantitative, and local.

Classification of Movements *per Se*

In respect of Substance there is no motion, because Substance has no contrary among things that are. Nor is there motion in respect of Relation: for it may happen that when one correlative changes, the other, although this does not itself change, is no longer applicable, so that in these cases the motion is accidental. Nor is there motion in respect of Agent and Patient – in fact there can never be motion of mover and moved, because there cannot be motion of motion or becoming of becoming or in general change of change.

For in the first place there are two senses in which motion of motion is conceivable. (1) The motion of which there is motion might be conceived as subject; e.g. a man is in motion because he changes from fair to dark. Can it be that in this sense motion grows hot or cold, or changes place, or increases or decreases? Impossible: for change is not a subject. Or (2) can there be motion of motion in the sense that some other subject changes from a change to another mode of being, as e.g. a man changes from falling ill to getting well? Even this is possible only in an accidental sense. For, whatever the subject may be, movement is change from one form to another. (And the same holds good of becoming and perishing, except that in these processes we have a change to a particular kind of opposite, while the other, motion, is a change to a different kind.) So, if there is to be motion of motion, that which is changing from health to sickness must simultaneously be changing from this very change to another. It is clear, then, that by the time that it has become sick, it must also have changed to whatever may be the other change concerned (for that it should be at rest, though logically possible, is excluded by the theory). Moreover this other can never be any casual change, but must be a change from something definite to some other definite thing. So in this case it must be the opposite change, viz. convalescence. It is only accidentally that there can be change of change, e.g. there is a change from remembering to forgetting only because the subject of this change changes at one time to knowledge, at another to ignorance.

170

In the second place, if there is to be change of change and becoming of becoming, we shall have an infinite regress. Thus if one of a series of changes is to be a change of change, the preceding change must also be so: e.g. if simple becoming was ever in process of becoming, then that which was becoming simple becoming was also in process of becoming, so that we should not yet have arrived at what was in process of simple becoming but only at what was already in process of becoming in process of becoming. And this again was sometime in process of becoming, so that even then we should not have arrived at what was in process of simple becoming. And since in an infinite series there is no first term, here there will be no first stage and therefore no following stage either. On this hypothesis, then, nothing can become or be moved or change.

Thirdly, if a thing is capable of any particular motion, it is also capable of the corresponding contrary motion or the corresponding coming to rest, and a thing that is capable of becoming is also capable of perishing: consequently, if there be becoming of becoming, that which is in process of becoming is in process of perishing at the very moment when it has reached the stage of becoming: since it cannot be in process of perishing when it is just beginning to become or after it has ceased to become: for that which is in process of perishing must be in existence.

Fourthly, there must be a substrate underlying all processes of becoming and changing. What can this be in the present case? It is either the body or the soul that undergoes alteration: what is it that correspondingly becomes motion or becoming?

171

And again what is the goal of their motion? It must be the motion or becoming of something from something to something else. But in what sense can this be so? For the becoming of learning cannot be learning: so neither can the becoming of becoming be becoming, nor can the becoming of any process be that process.

Finally, since there are three kinds of motion, the substratum and the goal of motion must be one or other of these, e.g. locomotion will have to be altered or to be locally moved.

To sum up, then, since everything that is moved is moved in one of three ways, either accidentally, or partially, or essentially, change can change only accidentally, as e.g. when a man who is being restored to health runs or learns: and accidental change we have long ago decided to leave out of account.

Since, then, motion can belong neither to Being nor to Relation nor to Agent and Patient, it remains that there can be motion only in respect of Quality, Quantity, and Place: for with each of these we have a pair of contraries. Motion in respect of Quality let us call alteration, a general designation that is used to include both contraries: and by Quality I do not here mean a property of substance (in that sense that which constitutes a specific distinction is a quality) but a passive quality in virtue of which a thing is said to be acted on or to be incapable of being acted on. Motion in respect of Quantity has no name that includes both contraries, but it is called increase or decrease according as one or the other is designated: that is to say motion in the direction of complete magnitude is increase, motion in the contrary direction is decrease. Motion in respect of Place has no name either general or particular: but we may designate it by the general name of locomotion, though strictly the term 'locomotion' is applicable to things that change their place only when they have not the power to come to a stand, and to things that do not move *themselves* locally.

Change within the same kind from a lesser to a greater or from a greater to a lesser degree is alteration: for it is motion either from a contrary or to a contrary, whether in an unqualified or in a qualified sense: for change to a lesser degree of a quality will be called change to the contrary of that quality and change to a greater degree of a quality will be regarded as change from the contrary of that quality to the quality itself. It makes no difference whether the change be qualified or unqualified, except that in the former case the contraries will have to be contrary to one another only in a qualified sense: and a thing's possessing a quality in a greater or in a lesser degree means the presence or absence in it of more or less of the opposite quality. It is now clear, then, that there are only these three kinds of motion.

The Unmovable

The term 'immovable' we apply in the first place to that which is absolutely incapable of being moved (just as we correspondingly apply the term invisible to sound); in the second place to that which is moved with difficulty after a long time or whose movement is slow at the start – in fact, what we describe as hard to move; and in the third place to that which is naturally designed for and capable of motion, but is not in motion when, where, and as it naturally would be so. This last is the only kind of immovable thing of which I use the term 'being at rest': for rest is contrary to motion, so that rest will be negation of motion in that which is capable of admitting motion.

The foregoing remarks are sufficient to explain the essential nature of motion and rest, the number of kinds of change, and the different varieties of motion.

The Meaning of 'Together', 'Apart', 'Touch', 'Intermediate', 'Successive', 'Contiguous', 'Continuous'

Let us now proceed to define the terms 'together' and 'apart', 'in contact', 'between', 'in succession', 'contiguous', and 'continuous', and to show in what circumstances each of these terms is naturally applicable.

Things are said to be together in place when they are in one place (in the strictest sense of the word 'place') and to be apart when they are in different places.

Things are said to be in contact when their extremities are together.

That which a changing thing, if it changes continuously in a natural manner, naturally reaches before it reaches that to which it changes last, is between. Thus 'between' implies the presence of at least three things: for in a process of change it is the contrary that is 'last': and a thing is moved continuously if it leaves no gap or only the smallest possible gap in the material – not in the time (for a gap in the time does not prevent things

174

having a 'between', while, on the other hand, there is nothing to prevent the highest note sounding immediately after the lowest) but in the material in which the motion takes place. This is manifestly true not only in local changes but in every other kind as well. (Now every change implies a pair of opposites, and opposites may be either contraries or contradictories; since then contradiction admits of no mean term, it is obvious that 'between' must imply a pair of contraries.) That is locally contrary which is most distant in a straight line: for the shortest line is definitely limited, and that which is definitely limited constitutes a measure.

A thing is 'in succession' when it is after the beginning in position or in form or in some other respect in which it is definitely so regarded, and when further there is nothing of the *same* kind as itself between it and that to which it is in succession, e.g. a line or lines if it is a line, a unit or units if it is a unit, a house if it is a house (there is nothing to prevent something of a *different* kind being between). For that which is in succession is in succession to a particular thing, and is something posterior: for one is not 'in succession' to two, nor is the first day of the month to the second: in each case the latter is 'in succession' to the former.

A thing that is in succession and touches is 'contiguous'. The 'continuous' is a subdivision of the contiguous: things are called continuous when the touching limits of each become one and the same and are, as the word implies, contained in each other: continuity is impossible if these extremities are two. This definition makes it plain that continuity belongs to things that

175

naturally in virtue of their mutual contact form a unity. And in whatever way that which holds them together is one, so too will the whole be one, e.g. by a rivet or glue or contact or organic union.

It is obvious that of these terms 'in succession' is first in order of analysis: for that which touches is necessarily in succession, but not everything that is in succession touches: and so succession is a property of things prior in definition, e.g. numbers, while contact is not. And if there is continuity there is necessarily contact, but if there is contact, that alone does not imply continuity: for the extremities of things may be '*together*' without necessarily being *one*: but they cannot be one without being necessarily together. So natural junction is last in coming to be: for the extremities must necessarily come into contact if they are to be naturally joined: but things that are in contact are not all naturally joined, while where there is no contact clearly there is no natural junction either. Hence, if as some say 'point' and 'unit' have an independent existence of their own, it is impossible for the two to be identical: for points can touch while units can only be in succession. Moreover, there can always be something between points (for all lines are intermediate between points), whereas it is not necessary that there should possibly be anything between units: for there can be nothing between the numbers one and two.

We have now defined what is meant by 'together' and 'apart', 'contact', 'between' and 'succession', 'contiguous' and 'continuous': and we have shown in what circumstances each of these terms is applicable.

The Unity and Diversity of Movements

There are many senses in which motion is said to be 'one': for we use the term 'one' in many senses.

Motion is one *generically* according to the different categories to which it may be assigned: thus any locomotion is one generically with any other locomotion, whereas alteration is different generically from locomotion.

Motion is one specifically when besides being one generically it also takes place in a species incapable of subdivision: e.g. colour has specific differences: therefore blackening and whitening differ specifically; but at all events every whitening will be specifically the same with every other whitening and every blackening with every other blackening. But whiteness is not further subdivided by specific differences: hence any whitening is specifically one with any other whitening. Where it happens that the genus is at the same time a species, it is clear that the motion will then in a sense be one specifically though not in an unqualified sense: learning is an example of this, knowledge being on the one hand a species of apprehension and on the other hand a genus including the various knowledges. A difficulty, however, may be raised as to whether a motion is specifically one when the same thing changes from the same to the same, e.g. when one point changes again and again from a particular place to a particular place: if this motion is specifically one, circular motion will be the same as rectilinear motion, and rolling the same as walking. But is not this difficulty removed by the principle already laid down that if that in which the motion takes place is specifically different (as in the present instance the circular path is specifically different from the straight) the motion itself is also different? We have explained, then, what is meant by saying that motion is one generically or one specifically.

Motion is one in an unqualified sense when it is one essentially or numerically: and the following distinctions will make clear what this kind of motion is. There are three classes of things in connexion with which we speak of motion, the 'that which', the 'that in which', and the 'that during which'. I mean that there must *be* something that is in motion, e.g. a man or gold, and it must be in motion *in* something, e.g. a place or an affection, and *during* something, for all motion takes place during a time. Of these three it is the thing in which the motion takes place that makes it one generically or specifically, it is the thing moved that makes the motion one in subject, and it is the time that makes it consecutive: but it is the three together that make it one without qualification: to effect this, that in which the motion takes place (the species) must be one and incapable of subdivision, that during which it takes place (the time) must be one and unintermittent, and that which is in motion must be one – not in an accidental sense (i.e. it must be one as the white that blackens is one or Coriscus who walks is one, not in the accidental sense in which Coriscus and white may be one), nor merely in virtue of community of nature (for there might be a case of two men being restored to health at the same time in the same way, e.g. from inflammation of the eye, yet this motion is not really one, but only specifically one).

Suppose, however, that Socrates undergoes an alteration specifically the same but at one time and again at another: in this case if it is possible for that which ceased to be again to come into being and remain numerically the same, then this motion too will be one: otherwise it will be the same but not one. And akin to this difficulty there is another; viz. is health one? and generally are the states and affections in bodies severally one in essence although (as is clear) the things that contain them are obviously in motion and in flux? Thus if a person's health at daybreak and at the present moment is one and the same, why should not this health be numerically one with that which he recovers after an interval? The same argument applies in each case. There is, however, we may answer, this difference: that if the states are two then it follows simply from this fact that the activities must also in point of number be two (for only that which is numerically one can give rise to an activity that is numerically one), but if the state is one, this is not in itself enough to make us regard the activity also as one: for when a man ceases walking, the walking no longer is, but it will again be if he begins to walk again. But, be this as it may, if in the above instance the health is one and the same, then it must be possible for that which is one and the same to come to be and to cease to be many times. However, these difficulties lie outside our present inquiry.

Since every motion is continuous, a motion that is one in an unqualified sense must (since every motion is divisible) be continuous, and a continuous motion must be one. There will not be continuity between any motion and any other indiscriminately any more than there is between any two things chosen at random in any other sphere: there can be continuity only when the extremities of the two things are one. Now some things have no extremities at all: and the extremities of others differ specifically although we give them the same name of 'end': how should e.g. the 'end' of a line and the 'end' of walking touch or come to be one? Motions that are not the same either specifically or generically may, it is true, be *consecutive* (e.g. a man may run and then at once fall ill of a fever), and again, in the torch-race we have consecutive but not continuous locomotion: for according to our definition there can be continuity only when the ends of the two things are one. Hence motions may be consecutive or successive in virtue of the time being continuous, but there can be continuity only in virtue of the motions themselves being continuous, that is when the end of each is one with the end of the other. Motion, therefore, that is in an unqualified sense continuous and one must be specifically the same, of one thing, and in one time. Unity is required in respect of time in order that there may be no interval of immobility, for where there is intermission of motion there must be rest, and a motion that includes intervals of rest will be not one but many, so that a motion that is interrupted by stationariness is not one or continuous, and it is so interrupted if there is an interval of time. And though of a motion that is not specifically one (even if the time is unintermittent) the time is one, the motion is specifically different, and so cannot really be one, for motion that is one must be specifically one, though motion that is specifically one is not necessarily one in an unqualified sense.

We have now explained what we mean when we call a motion one without qualification.

Further, a motion is also said to be one generically, specifically, or essentially when it is complete, just as in other cases completeness and wholeness are characteristics of what is one: and sometimes a motion even if incomplete is said to be one, provided only that it is continuous.

And besides the cases already mentioned there is another in which a motion is said to be one, viz. when it is regular: for in a sense a motion that is irregular is not regarded as one, that title belonging rather to that which is regular, as a straight line is regular, the irregular being as such divisible. But the difference would seem to be one of degree. In every kind of motion we may have regularity or irregularity: thus there may be regular alteration, and locomotion in a regular path, e.g. in a circle or on a straight line, and it is the same with regard to increase and decrease. The difference that makes a motion irregular is sometimes to be found in its path: thus a motion cannot be regular if its path is an irregular magnitude, e.g. a broken line, a spiral, or any other magnitude that is not such that any part of it taken at random fits on to any other that may be chosen. Sometimes it is found neither in the place nor in the time nor in the goal but in the manner of the motion: for in some cases the motion is differentiated by quickness and slowness: thus if its velocity is uniform a motion is regular, if not it is irregular. So quickness and slowness are not species of motion nor do they constitute specific differences of motion, because this distinction occurs in connexion with all the distinct species of motion. The same is true of heaviness and lightness when they refer to the same thing: e.g. they do not specifically distinguish earth from itself or fire from itself. Irregular motion, therefore, while in virtue of being continuous it is one, is so in a lesser degree, as is the case with locomotion in a broken line: and a lesser degree of something always means an admixture of its contrary. And since every motion that is one can be both regular and irregular, motions that are consecutive but not specifically the same cannot be one and continuous: for how should a motion composed of alteration and locomotion be regular? If a motion is to be regular its parts ought to fit one another.

Contrariety of Movement

We have further to determine what motions are contrary to each other, and to determine similarly how it is with rest. And we have first to decide whether contrary motions are motions respectively from and to the same thing, e.g. a motion from health and a motion to health (where the opposition, it would seem, is of the same kind as that between coming to be and ceasing to be); or motions respectively from contraries, e.g. a motion from health and a motion from disease; or motions respectively to contraries, e.g. a motion to health and a motion to disease; or motions respectively from a contrary and to the opposite contrary, e.g. a motion from health and a motion to disease; or motions respectively from a contrary to the opposite contrary and from the latter to the former, e.g. a motion from health to disease and a motion from disease to health: for motions must be contrary to one another in one or more of these ways, as there is no other way in which they can be opposed.

Now motions respectively from a contrary and to the opposite contrary, e.g. a motion from health and a motion to disease, are not contrary motions: for they are one and the same. (Yet their essence is not the same, just as changing from health is different from changing to disease.) Nor are motions respectively from a contrary and from the opposite contrary contrary motions, for a motion from a contrary is at the same time a motion to a contrary or to an intermediate (of this, however, we shall speak later), but changing to a contrary rather than changing from a contrary would seem to be the cause of the contrariety of motions, the latter being the loss, the former the gain, of contrariness. Moreover, each several motion takes its name rather from the goal than from the starting-point of change, e.g. motion to health we call convalescence, motion to disease sickening. Thus we are left with motions respectively to contraries, and motions respectively to contraries from the opposite contraries. Now it would seem that motions to contraries are at the same time motions from contraries (though their essence may not be the same; 'to health' is distinct, I mean, from 'from disease', and 'from health' from 'to disease').

Since then change differs from motion (motion being change from a particular subject to a particular subject), it follows that contrary motions are motions respectively from a contrary to the opposite contrary and from the latter to the former, e.g. a motion from health to disease and a motion from disease to health. Moreover, the consideration of particular

examples will also show what kinds of processes are generally
recognized as contrary: thus falling ill is regarded as contrary
to recovering one's health, these processes having contrary
goals, and being taught as contrary to being led into error
by another, it being possible to acquire error, like knowledge,
either by one's own agency or by that of another. Similarly
we have upward locomotion and downward locomotion,
which are contrary lengthwise, locomotion to the right and
locomotion to the left, which are contrary breadthwise, and
forward locomotion and backward locomotion, which too
are contraries.

On the other hand, a process simply to a contrary, e.g. that
denoted by the expression 'becoming white', where no starting-
point is specified, is a change but not a motion. And in all
cases of a thing that has no contrary we have as contraries
change from and change to the same thing. Thus coming to be
is contrary to ceasing to be, and losing to gaining. But these
are changes and not motions. And wherever a pair of contraries
admit of an intermediate, motions to that intermediate must be
held to be in a sense motions to one or other of the contraries:
for the intermediate serves as a contrary for the purposes of the
motion, in whichever direction the change may be, e.g. grey in
a motion from grey to white takes the place of black as starting-
point, in a motion from white to grey it takes the place of black
as goal, and in a motion from black to grey it takes the place
of white as goal: for the middle is opposed in a sense to either
of the extremes, as has been said above. Thus we see that two
motions are contrary to each other only when one is a motion

from a contrary to the opposite contrary and the other is a motion from the latter to the former.

Contrariety of Movement and Rest

But since a motion appears to have contrary to it not only another motion but also a state of rest, we must determine how this is so. A motion has for its contrary in the strict sense of the term another motion, but it also has for an opposite a state of rest (for rest is the privation of motion and the privation of anything may be called its contrary), and motion of one kind has for its opposite rest of that kind, e.g. local motion has local rest. This statement, however, needs further qualification: there remains the question, is the opposite of remaining at a particular place motion from or motion to that place? It is surely clear that since there are two subjects between which motion takes place, motion from one of these (A) to its contrary (B) has for its opposite remaining in A, while the reverse motion has for its opposite remaining in B. At the same time these two are also contrary to each other: for it would be absurd to suppose that there are contrary motions and not opposite states of rest. States of rest in contraries *are* opposed. To take an example, a state of rest in health is (1) contrary to a state of rest in disease, and (2) the motion to which it is contrary is that from health to disease. For (2) it would be absurd that its contrary motion should be that from disease to health, since motion to that in which a thing is at rest is rather a coming to rest, the coming to rest being found to come into being simultaneously with the motion; and one of these two motions it must be. And (1) rest in *whiteness* is of course not contrary to rest in health.

Of all things that have no contraries there are opposite *changes* (viz. change from the thing and change to the thing, e.g. change from being and change to being), but no *motion*. So, too, of such things there is no remaining though there is absence of change. Should there be a particular subject, absence of change in its being will be contrary to absence of change in its not-being. And here a difficulty may be raised: if not-being is not a particular something, what is it, it may be asked, that is contrary to absence of change in a thing's being? and is this absence of change a state of rest? If it is, then either it is not true that every state of rest is contrary to a motion or else coming to be and ceasing to be are motion. It is clear then that, since we exclude these from among motions, we must not say that this absence of change is a state of rest: we must say that it is similar to a state of rest and call it absence of change. And it will have for its contrary either nothing or absence of change in the thing's not-being, or the ceasing to be of the thing: for such ceasing to be is change from it and the thing's coming to be is change to it.

Contrariety of Natural and Unnatural Movement or Rest

Again, a further difficulty may be raised. How is it, it may be asked, that whereas in local change both remaining and moving may be natural or unnatural, in the other changes this is not so? e.g. alteration is not now natural and now unnatural, for convalescence is no more natural or unnatural than falling ill, whitening no more natural or unnatural than blackening; so, too, with increase and decrease: these are not contrary to each other in the sense that either of them is natural while the other is unnatural, nor is one increase contrary to another in this sense; and the same account may be given of becoming and perishing: it is not true that becoming is natural and perishing unnatural (for growing old is natural), nor do we observe one becoming to be natural and another unnatural. We answer that if what happens under violence is unnatural, then violent perishing is unnatural and as such contrary to natural perishing. Are there then also some becomings that are violent and not the result of natural necessity, and are therefore contrary to natural becomings, and violent increases and decreases, e.g. the rapid growth to maturity of profligates and the rapid ripening of seeds even when not packed close in the earth? And how is it with alterations? Surely just the same: we may say that some alterations are violent while others arc natural, e.g. patients alter naturally or unnaturally according as they throw off fevers on the critical days or not. But, it may be objected, then we shall have perishings contrary to one another, not to becoming. Certainly: and why should not this in a sense be so? Thus it is so if one perishing is pleasant and another painful: and so one perishing will be contrary to another not in an unqualified sense, but in so far as one has this quality and the other that.

Now motions and states of rest universally exhibit contrariety in the manner described above, e.g. upward motion and rest above are respectively contrary to downward motion and rest below, these being instances of local contrariety; and upward locomotion belongs naturally to fire and downward to earth, i.e. the locomotions of the two are contrary to each other. And again, fire moves up naturally and down unnaturally: and its natural motion is certainly contrary to its unnatural motion. Similarly with remaining: remaining above is contrary to motion from above downwards, and to earth this remaining comes unnaturally, this motion naturally. So the unnatural remaining of a thing is contrary to its natural motion, just as we find a similar contrariety in the motion of the same thing: one of its motions, the upward or the downward, will be natural, the other unnatural.

Here, however, the question arises, has every state of rest that is not permanent a becoming, and is this becoming a coming to a standstill? If so, there must be a becoming of that which is at rest unnaturally, e.g. of earth at rest above: and therefore this earth during the time that it was being carried violently upward was coming to a standstill. But whereas the velocity of that which comes to a standstill seems always to increase, the velocity of that which is carried violently seems always to decrease: so it will *be* in a state of rest without having *become* so. Moreover 'coming to a standstill' is generally recognized to be identical or at least concomitant with the locomotion of a thing to its proper place.

There is also another difficulty involved in the view that remaining in a particular place is contrary to motion from that place. For when a thing is moving from or discarding something, it still appears to have that which is being discarded, so that if a state of rest is itself contrary to the motion from the state of rest to its contrary, the contraries rest and motion will be simultaneously predicable of the same thing. May we not say, however, that in so far as the thing is still stationary it is in a state of rest in a qualified sense? For, in fact, whenever a thing is in motion, part of it is at the starting-point while part is at the goal to which it is changing: and consequently a motion finds its true contrary rather in another motion than in a state of rest.

With regard to motion and rest, then, we have now explained in what sense each of them is one and under what conditions they exhibit contrariety.

[With regard to coming to a standstill the question may be raised whether there is an opposite state of rest to unnatural as well as to natural motions. It would be absurd if this were not the case: for a thing may remain still merely under violence: thus we shall have a thing being in a non-permanent state of rest without having become so. But it is clear that it must be the case: for just as there is unnatural motion, so, too, a thing may be in an unnatural state of rest. Further, some things have a natural and an unnatural motion, e.g. fire has a natural upward motion and an unnatural downward motion: is it, then, this unnatural downward motion or is it the natural downward motion of earth that is contrary to the natural upward motion? Surely it is clear that both are contrary to it though not in the same sense: the natural motion of earth is contrary inasmuch as the motion of fire is also natural, whereas the upward motion of fire as being natural is contrary to the downward motion of fire as being unnatural. The same is true of the corresponding cases of remaining. But there would seem to be a sense in which a state of rest and a motion are opposites.]

Book VI

Every Continuum Consists of Continuous and Divisible Parts

Now if the terms 'continuous', 'in contact', and 'in succession' are understood as defined above – things being 'continuous' if their extremities are one, 'in contact' if their extremities are together, and 'in succession' if there is nothing of their own kind intermediate between them – nothing that is continuous can be composed of indivisibles: e.g. a line cannot be composed of points, the line being continuous and the point indivisible. For the extremities of two points can neither be *one* (since of an indivisible there can be no extremity as distinct from some other part) nor *together* (since that which has no parts can have no extremity, the extremity and the thing of which it is the extremity being distinct).

Moreover, if that which is continuous is composed of points, these points must be either *continuous* or *in contact* with one another: and the same reasoning applies in the case of all indivisibles. Now for the reason given above they cannot be continuous: and one thing can be in contact with another only if whole is in contact with whole or part with part or part with whole. But since indivisibles have no parts, they must be in contact with one another as whole with whole. And if they are in contact with one another as whole with whole, they will not be continuous: for that which is continuous has distinct parts: and these parts into which it is divisible are different in this way, i.e. spatially separate.

Nor, again, can a point be *in succession* to a point or a moment to a moment in such a way that length can be composed of points or time of moments: for things are in succession if there is nothing of their own kind intermediate between them, whereas that which is intermediate between points is always a line and that which is intermediate between moments is always a period of time.

Again, if length and time could thus be composed of indivisibles, they could be divided into indivisibles, since each is divisible into the parts of which it is composed. But, as we saw, no continuous thing is divisible into things without parts. Nor can there be anything of any other kind intermediate between the parts or between the moments: for if there could be any such thing it is clear that it must be either indivisible or divisible, and if it is divisible, it must be divisible either into indivisibles or into divisibles that are infinitely divisible, in which case it is continuous.

Moreover, it is plain that everything continuous is divisible into divisibles that are infinitely divisible: for if it were divisible into indivisibles, we should have an indivisible in contact with an indivisible, since the extremities of things that are continuous with one another are one and are in contact.

The same reasoning applies equally to magnitude, to time, and to motion: either all of these are composed of indivisibles and are divisible into indivisibles, or none. This may be made clear as follows. If a magnitude is composed of indivisibles, the motion over that magnitude must be composed of corresponding indivisible motions: e.g. if the magnitude ABΓ is composed of the indivisibles A, B, Γ, each corresponding part of the motion ΔEZ of Ω over ABΓ is indivisible. Therefore, since where there is motion there must be something that is in motion, and where there is something in motion there must be motion, therefore the being-moved will also be composed of indivisibles. So Ω traversed A when its motion was Δ, B when its motion was E, and Γ similarly when its motion was Z. Now a thing that is in motion from one place to another cannot at the moment when it was in motion both be in motion and at the same time have completed its motion at the place to which it was in motion: e.g. if a man is walking to Thebes, he cannot be walking to Thebes and at the same time have completed his walk to Thebes: and as we saw, Ω traverses the partless section A in virtue of the presence of the motion Δ. Consequently, if Ω actually passed through A *after* being in process of passing through, the motion must be divisible: for at the time when Ω was passing through, it neither was at rest nor had completed its passage but was in an intermediate state: while if it is passing through and has completed its passage *at the same moment*, then that which is walking will at the moment when it is walking have completed its walk and will be in the place to which it is walking; that is to say, it will have completed its motion at the place to which it is in motion. And if a thing is in motion over the ABΓ and its motion is the three Δ, E, and Z, and if it is not in motion at all over the partless section A but has competed its motion over it, then the motion will consist not of motions but of starts, and will take place by a thing's having completed a motion without being in motion: for on this assumption it has completed its passage through A without passing through it. So it will be possible for a thing to have completed a walk without ever walking: for on this assumption it has completed a walk over a particular distance without walking over that distance. Since, then, everything must be either at rest or in motion, and Ω is therefore at rest in each of the sections A, B, and Γ, it follows that, a thing, can be continuously at rest and at the same time in motion: for, as we saw, Ω is in motion over the whole ABΓ and at rest in any part (and consequently in the whole) of it. Moreover, if the indivisibles composing ΔEZ are motions, it would be possible for a thing in spite of the presence in it of motion to be not in motion but at rest, while if they are not motions, it would be possible for motion to be composed of something other than motions.

And if length and motion are thus indivisible, it is neither more nor less necessary that time also be similarly indivisible, that is to say be composed of indivisible moments: for if the whole distance is divisible and an equal velocity will cause a thing to pass through less of it in less time, the time must also be divisible, and conversely, if the time in which a thing is carried over the section A is divisible, this section A must also be divisible.

And since every magnitude is divisible into magnitudes – for we have shown that it is impossible for anything continuous to be composed of indivisible parts, and every magnitude is continuous – it necessarily follows that the quicker of two things traverses a greater magnitude in an equal time, an equal magnitude in less time, and a greater magnitude in less time, in conformity with the definition sometimes given of 'the quicker'. Suppose that A is quicker than B. Now since of two things that which changes sooner is quicker, in the time ZH, in which A has changed from Γ to Δ, B will not yet have arrived at Δ but will be short of it: so that in an equal time the quicker will pass over a greater magnitude. More than this, it will pass over a greater magnitude in less time: for in the time in which A has arrived at Δ, B being the slower has arrived, let us say, at E. Then since A has occupied the whole time ZH in arriving at Δ, it will have arrived at Θ in less time than this, say ZK. Now the magnitude ΓΘ that A has passed over is greater than the magnitude ΓE, and the time ZK is less than the whole time ZH: so that the quicker will pass over a greater magnitude in less time. And from this it is also clear that the quicker will pass over an equal magnitude in less time than the slower. For since it passes over the greater magnitude in less time than the slower, and (regarded by itself) passes over ΛM the greater in more time than ΛΞ the lesser, the time ΠP in which it passes over ΛM will be more than the time ΠΣ in which it passes over ΛΞ: so that, the time ΠΡ being less than the time ΠΧ in which the slower passes over ΛΞ, the time ΠΣ will also be less than the time ΠΧ: for it is less than the time ΠΡ, and that which is less than something else that is less than a thing is also itself less than that thing. Hence it follows that the quicker will traverse an equal magnitude in less time than the slower.

Again, since the motion of anything must always occupy either an equal time or less or more time in comparison with that of another thing, and since, whereas a thing is slower if its motion occupies more time and of equal velocity if its motion occupies an equal time, the quicker is neither of equal velocity nor slower, it follows that the motion of the quicker can occupy neither an equal time nor more time. It can only be, then, that it occupies less time, and thus we get the necessary consequence that the quicker will pass over an equal magnitude (as well as a greater) in less time than the slower. And since every motion is in time and a motion may occupy any time, and the motion of everything that is in motion may be either quicker or slower, both quicker motion and slower motion may occupy any time: and this being so, it necessarily follows that time also is continuous. By continuous I mean that which is divisible into divisibles that are infinitely divisible: and if we take this as the definition of continuous, it follows necessarily that time is continuous. For since it has been shown that the quicker will pass over an equal magnitude in less time than the slower, suppose that A is quicker and B slower, and that the slower has traversed the magnitude ΓΔ in the time ΖΗ. Now it is clear that the quicker will traverse the same magnitude in less time

than this: let us say in the time ΖΘ. Again, since the quicker has
passed over the whole ΓΔ in the time ΖΘ, the slower will in the
same time pass over ΓΚ, say, which is less than ΓΔ. And since
Β, the slower, has passed over ΓΚ in the time ΖΘ, the quicker
will pass over it in less time: so that the time ΖΘ will again be
divided. And if this is divided the magnitude ΓΚ will also be
divided just as ΓΔ was: and again, if the magnitude is divided,
the time will also be divided. And we can carry on this process
for ever, taking the slower after the quicker and the quicker after
the slower alternately, and using what has been demonstrated
at each stage as a new point of departure: for the quicker will
divide the time and the slower will divide the length. If, then,
this alternation always holds good, and at every turn involves a
division, it is evident that all time must be continuous. And at
the same time it is clear that all magnitude is also continuous;
for the divisions of which time and magnitude respectively are
susceptible are the same and equal.

Moreover, the current popular arguments make it plain that,
if time is continuous, magnitude is continuous also, inasmuch as a
thing passes over half a given magnitude in half the time taken to
cover the whole: in fact without qualification it passes over a less
magnitude in less time; for the divisions of time and of magnitude
will be the same. And if either is infinite, so is the other, and the
one is so in the same way as the other; i.e. if time is infinite in
respect of its extremities, length is also infinite in respect of its
extremities: if time is infinite in respect of divisibility, length is
also infinite in respect of divisibility: and if time is infinite in both
respects, magnitude is also infinite in both respects.

Hence Zeno's argument makes a false assumption in asserting that it is impossible for a thing to pass over or severally to come in contact with infinite things in a finite time. For there are two senses in which length and time and generally anything continuous are called 'infinite': they are called so either in respect of divisibility or in respect of their extremities. So while a thing in a finite time cannot come in contact with things quantitatively infinite, it can come in contact with things infinite in respect of divisibility: for in this sense the time itself is also infinite: and so we find that the time occupied by the passage over the infinite is not a finite but an infinite time, and the contact with the infinites is made by means of moments not finite but infinite in number.

The passage over the infinite, then, cannot occupy a finite time, and the passage over the finite cannot occupy an infinite time: if the time is infinite the magnitude must be infinite also, and if the magnitude is infinite, so also is the time. This may be shown as follows. Let AB be a finite magnitude, and let us suppose that it is traversed in infinite time Γ, and let a finite period ΓΔ of the time be taken. Now in this period the thing in motion will pass over a certain segment of the magnitude: let BE be the segment that it has thus passed over. (This will be either an exact measure of AB or less or greater than an exact measure: it makes no difference which it is.) Then, since a magnitude equal to BE will always be passed over in an equal time, and BE measures the whole magnitude, the whole time occupied in passing over AB will be finite: for it will be divisible into periods equal in number to the segments into which the magnitude is divisible. Moreover, if it is the case that infinite time is not occupied in passing over every magnitude, but it is possible to pass over some magnitude, say BE, in a finite time, and if this BE measures the whole of which it is a part, and if an equal magnitude is passed over in an equal time, then it follows that the time like the magnitude is finite. That infinite time will not be occupied in passing over BE is evident if the time be taken as limited in one direction: for as the part will be passed over in less time than the whole, the time occupied in traversing this part must be finite, the limit in one direction being given. The same reasoning will also show the falsity of the assumption that infinite length can be traversed in a finite time. It is evident, then, from what has been said that neither a line nor a surface nor in fact anything continuous can be indivisible.

This conclusion follows not only from the present argument but from the consideration that the opposite assumption implies the divisibility of the indivisible. For since the distinction of quicker and slower may apply to motions occupying any period of time and in an equal time the quicker passes over a greater length, it may happen that it will pass over a length twice, or one and a half times, as great as that passed over by the slower: for their respective velocities may stand to one another in this proportion. Suppose, then, that the quicker has in the same time been carried over a length one and a half times as great as that traversed by the slower, and that the respective magnitudes are divided, that of the quicker, the magnitude ABΓΔ into three indivisibles, and that of the slower into the two indivisibles EZ, ZH. Then the time may also be divided into three indivisibles, for an equal magnitude will be passed over in an equal time. Suppose then that it is thus divided into KA, ΛM, MN. Again, since in the same time the slower has been carried over EZ, ZH, the time may also be similarly divided into two. Thus the indivisible will be divisible, and that which has no parts will be passed over not in an indivisible but in a greater time. It is evident, therefore, that nothing continuous is without parts.

A Moment Is Indivisible and Nothing Is Moved, or Rests, in a Moment

The present also is necessarily indivisible – the present, that is, not in the sense in which the word is applied to one thing in virtue of another, but in its proper and primary sense; in which sense it is inherent in all time. For the present is something that is an extremity of the past (no part of the future being on this side of it) and also of the future (no part of the past being on the other side of it): it is, as we have said, a limit of both. And if it is once shown that it is essentially of this character and one and the same, it will at once be evident also that it is indivisible.

Now the present that is the extremity of both times must be one and the same: for if each extremity were different, the one could not be in succession to the other, because nothing continuous can be composed of things having no parts: and if the one is apart from the other, there will be time intermediate between them, because everything continuous is such that there is something intermediate between its limits and described by the same name as itself. But if the intermediate thing is time, it will be divisible: for all time has been shown to be divisible. Thus on this assumption the present is divisible. But if the present is divisible, there will be part of the past in the future and part of the future in the past: for past time will be marked off from future time at the actual point of division. Also the present will be a present not in the proper sense but in virtue of something else: for the division which yields it will not be a division proper. Furthermore, there will be a part of the present that is past and a part that is future, and it will not always be the same part that is past or future: in fact one and the same present will not be simultaneous: for the time may be divided at many points. If, therefore, the present cannot possibly have these characteristics, it follows that it must be the same present that belongs to each of the two times. But if this is so it is evident that the present is also indivisible: for if it is divisible it will be involved in the same implications as before. It is clear, then, from what has been said that time contains something indivisible, and this is what we call a present.

We will now show that nothing can be in motion in a present. For if this is possible, there can be both quicker and slower motion in the present. Suppose then that in the present N the quicker has traversed the distance AB. That being so, the slower will in the same present traverse a distance less than AB, say AΓ. But since the slower will have occupied the whole present in traversing AΓ, the quicker will occupy less than this in traversing it. Thus we shall have a division of the present, whereas we found it to be indivisible. It is impossible, therefore, for anything to be in motion in a present.

Nor can anything be at rest in a present: for, as we were saying, that only can be at rest which is naturally designed to be in motion but is not in motion when, where, or as it would naturally be so: since, therefore, nothing is naturally designed to be in motion in a present, it is clear that nothing can be at rest in a present either.

Moreover, inasmuch as it is the same present that belongs to both the tirnes, and it is possible for a thing to be in motion throughout one time and to be at rest throughout the other, and that which is in motion or at rest for the whole of a time will be in motion or at rest as the case may be in any part of it in which it is naturally designed to be in motion or at rest: this being so, the assumption that there can be motion or rest in a present will carry with it the implication that the same thing can at the same time be at rest and in motion: for both the times have the same extremity, viz. the present.

Again, when we say that a thing is at rest, we imply that its condition in whole and in part is at the time of speaking uniform with what it was previously: but the present contains no 'previously': consequently, there can be no rest in it.

It follows then that the motion of that which is in motion and the rest of that which is at rest must occupy time.

Whatever Is Moved Is Divisible

Further, everything that changes must be divisible. For since every change is from something to something, and when a thing is at the goal of its change it is no longer changing, and when both it itself and all its parts are at the starting-point of its change it is not changing (for that which is in whole and in part in an unvarying condition is not in a state of change); it follows, therefore, that part of that which is changing must be at the starting-point and part at the goal: for as a whole it cannot be in both or in neither. (Here by 'goal of change' I mean that which comes first in the process of change: e.g. in a process of change from white the goal in question will be grey, not black: for it is not necessary that that which is changing should be at either of the extremes.) It is evident, therefore, that everything that changes must be divisible.

Classification of Movement

Now motion is divisible in two senses. In the first place it is divisible in virtue of the time that it occupies. In the second place it is divisible according to the motions of the several parts of that which is in motion: e.g. if the whole AΓ is in motion, there will be a motion of AB and a motion of BΓ. That being so, let ΔE be the motion of the part AB and EZ the motion of the part BΓ. Then the whole ΔZ must be the motion of AΓ: for ΔZ must constitute the motion of AΓ inasmuch as ΔE and EZ severally constitute the motions of each of its parts. But the motion of a thing can never be constituted by the motion of something else: consequently the whole motion is the motion of the whole magnitude.

Again, since every motion is a motion of something, and the whole motion ΔZ is not the motion of either of the parts (for each of the parts ΔE, EZ is the motion of one of the parts AB, $B\Gamma$) or of anything else (for, the whole motion being the motion of a whole, the parts of the motion are the motions of the parts of that whole: and the parts of ΔZ are the motions of AB, $B\Gamma$ and of nothing else: for, as we saw, a motion that is one cannot be the motion of more things than one): since this is so, the whole motion will be the motion of the magnitude $AB\Gamma$.

Again, if there is a motion of the whole other than ΔZ, say ΘI, the motion of each of the parts may be subtracted from it: and these motions will be equal to ΔE, EZ respectively: for the motion of that which is one must be one. So if the whole motion ΘI may be divided into the motions of the parts, ΘI will be equal to ΔZ: if on the other hand there is any remainder, say KI, this will be a motion of nothing: for it can be the motion neither of the whole nor of the parts (as the motion of that which is one must be one) nor of anything else: for a motion that is continuous must be the motion of things that are continuous. And the same result follows if the division of ΘI reveals a surplus on the side of the motions of the parts. Consequently, if this is impossible, the whole motion must be the same as and equal to ΔZ.

The Time, the Movement, the Being-in-Motion, the Moving Body, and the Sphere of Movement, Are All Similarly Divided

This then is what is meant by the division of motion according to the motions of the parts: and it must be applicable to everything that is divisible into parts.

Motion is also susceptible of another kind of division, that according to time. For since all motion is in time and all time is divisible, and in less time the motion is less, it follows that every motion must be divisible according to time. And since everything that is in motion is in motion in a certain sphere and for a certain time and has a motion belonging to it, it follows that the time, the motion, the being-in-motion, the thing that is in motion, and the sphere of the motion must all be susceptible of the same divisions (though spheres of motion are not all divisible in a like manner: thus quantity is essentially, quality accidentally divisible). For suppose that A is the time occupied by the motion B. Then if all the time has been occupied by the whole motion, it will take less of the motion to occupy half the time, less again to occupy a further subdivision of the time, and so on to infinity. Again, the time will be divisible similarly to the motion: for if the whole motion occupies all the time half the motion will occupy half the time, and less of the motion again will occupy less of the time.

In the same way the being-in-motion will also be divisible. For let Γ be the whole being-in-motion. Then the being-in-motion that corresponds to half the motion

204

will be less than the whole being-in-motion, that which corresponds to a quarter of the motion will be less again, and so on to infinity. Moreover by setting out successively the beingin-motion corresponding to each of the two motions ΔΓ (say) and ΓΕ, we may argue that the whole being-in-motion will correspond to the whole motion (for if it were some other being-in-motion that corresponded to the whole motion, there would be more than one being-in-motion corresponding to the same motion), the argument being the same as that whereby we showed that the motion of a thing is divisible into the motions of the parts of the thing: for if we take separately the being-in-motion corresponding to each of the two motions, we shall see that the whole being-in-motion is continuous.

The same reasoning will show the divisibility of the length, and in fact of everything that forms a sphere of change (though some of these are only accidentally divisible because that which changes is so): for the division of one term will involve the division of all. So, too, in the matter of their being finite or infinite, they will all alike be either the one or the other. And we now see that in most cases the fact that all the terms are divisible or infinite is a direct consequence of the fact that the thing that changes is divisible or infinite: for the attributes 'divisible' and 'infinite' belong in the first instance to the thing that changes. That divisibility does so we have already shown; that infinity does so will be made clear in what follows.

Whatever Has Changed Is, as Soon as It Has Changed, in that to Which It Has Changed

Since everything that changes changes from something to something, that which has changed must at the moment when it has first changed be in that to which it has changed. For that which changes retires from or leaves that from which it changes: and leaving, if not identical with changing, is at any rate a consequence of it. And if leaving is a consequence of changing, having left is a consequence of having changed: for there is a like relation between the two in each case.

One kind of change, then, being change in a relation of contradiction, where a thing has changed from not-being to being it has left not-being. Therefore it will be in being: for everything must either be or not be. It is evident, then, that in contradictory change that which has changed must be in that to which it has changed. And if this is true in this kind of change, it will be true in all other kinds as well: for in this matter what holds good in the case of one will hold good likewise in the case of the rest.

Moreover, if we take each kind of change separately, the truth of our conclusion will be equally evident, on the ground that that which has changed must be somewhere or in something. For, since it has left that from which it has changed and must be somewhere, it must be either in that to which it has changed or in something else. If, then, that which has changed to B is in something other than B, say Γ, it must again be changing from Γ to B: for it cannot be assumed that there is no interval

206

between Γ and B, since change is continuous. Thus we have the result that the thing that has changed, at the moment when it has changed, is changing to that to which it has changed, which is impossible: that which has changed, therefore, must be in that to which it has changed. So it is evident likewise that that which has come to be, at the moment when it has come to be, will *be*, and that which has ceased to be will *not-be*: for what we have said applies universally to every kind of change, and its truth is most obvious in the case of contradictory change. It is clear, then, that that which has changed, at the moment when it has first changed, is in that to which it has changed.

That in Which (Directly) It Has Changed Is Indivisible

We will now show that the 'primary when' in which that which has changed effected the completion of its change must be indivisible, where by 'primary' I mean possessing the characteristics in question of itself and not in virtue of the possession of them by something else belonging to it. For let AΓ be divisible, and let it be divided at B. If then the *completion* of change has been effected in AB or again in BΓ, AΓ cannot be the primary thing in which the completion of change has been effected. If, on the other hand, it has been changing in both AB and BΓ (for it must either have changed or be changing in each of them), it must have been changing in the whole AΓ: but our assumption was that AΓ contains only the completion of the change. It is equally impossible to suppose that one part of AΓ contains the process and the other the completion of

207

the change: for then we shall have something prior to what is primary. So that in which the completion of change has been effected must be indivisible. It is also evident, therefore, that that in which that which has ceased to be has ceased to be and that in which that which has come to be has come to be are indivisible.

In Change There Is a Last but No First Element

But there are two senses of the expression 'the primary when in which something has changed'. On the one hand it may mean the primary when containing the *completion* of the process of change – the moment when it is correct to say 'it has changed': on the other hand it may mean the primary when containing the *beginning* of the process of change. Now the primary when that has reference to the *end* of the change is something really existent: for a change may really be completed, and there is such a thing as an end of change, which we have in fact shown to be indivisible because it is a limit. But that which has reference to the beginning is not existent at all: for there is no such thing as a beginning of a process of change, and the time occupied by the change does not contain any primary when in which the change began. For suppose that AΔ is such a primary when. Then it cannot be indivisible: for, if it were, the moment immediately preceding the change and the moment in which the change begins would be consecutive (and moments cannot be consecutive). Again, if the changing thing is at rest in the whole preceding time ΓA (for we may suppose that it is at rest), it is at rest in A also: so if AΔ is without parts, it will

208

simultaneously be at rest and have changed: for it is at rest in A and has changed in Δ. Since then $A\Delta$ is not without parts, it must be divisible, and the changing thing must have changed in every part of it (for if it has changed in neither of the two parts into which $A\Delta$ is divided, it has not changed in the whole either: if, on the other hand, it is in process of change in both parts, it is likewise in process of change in the whole: and if, again, it has changed in one of the two parts, the whole is not the primary when in which it has changed: it must therefore have changed in every part). It is evident, then, that with reference to the beginning of change there is no primary when in which change has been effected: for the divisions are infinite.

So, too, of that which has changed there is no primary part that has changed. For suppose that of ΔE the primary part that has changed is ΔZ (everything that changes having been shown to be divisible): and let ΘI be the time in which ΔZ has changed. If, then, in the whole time ΔZ has changed, in half the time there will be a part that has changed, less than and therefore prior to ΔZ: and again there will be another part prior to this, and yet another, and so on to infinity. Thus of that which changes there cannot be any primary part that has changed. It is evident, then, from what has been said, that neither of that which changes nor of the time in which it changes is there any primary part.

With regard, however, to the actual subject of change – that is to say that in respect of which a thing changes – there is a difference to be observed. For in a process of change we may distinguish three terms – that which changes, that in which it changes, and the actual subject of change, e.g. the man, the time, and the fair complexion. Of these the man and the time are divisible: but with the fair complexion it is otherwise (though they are all divisible accidentally, for that in which the fair complexion or any other quality is an accident is divisible). For of actual subjects of change it will be seen that those which are classed as essentially, not accidentally, divisible have no primary part. Take the case of magnitudes: let AB be a magnitude, and suppose that it has moved from B to a primary 'where' Γ. Then if BΓ is taken to be indivisible, two things without parts will have to be contiguous (which is impossible): if on the other hand it is taken to be divisible, there will be something prior to Γ to which the magnitude has changed, and something else again prior to that, and so on to infinity, because the process of division may be continued without end. Thus there can be no primary 'where' to which a thing has changed. And if we take the case of quantitative change, we shall get a like result, for here too the change is in something continuous. It is evident, then, that only in qualitative motion can there be anything essentially indivisible.

In Whatever Time a Thing Changes (Directly), It Changes in Any Part of that Time

Now everything that changes changes in time, and that in two senses: for the time in which a thing is said to change may be the primary time, or on the other hand it may have an extended reference, as e.g. when we say that a thing changes in a particular year because it changes in a particular day. That being so, that which changes must be changing in any part of the primary time in which it changes. This is clear from our definition of 'primary', in which the word is said to express just this: it may also, however, be made evident by the following argument. Let XP be the primary time in which that which is in motion is in motion: and (as all time is divisible) let it be divided at K. Now in the time XK it either is in motion or is not in motion, and the same is likewise true of the time KP. Then if it is in motion in neither of the two parts, it will be at rest in the whole: for it is impossible that it should be in motion in a time in no part of which it is in motion. If on the other hand it is in motion in only one of the two parts of the time, XP cannot be the primary time in which it is in motion: for its motion will have reference to a time other than XP. It must, then, have been in motion in any part of XP.

Whatever Changes Has Changed Before, and Whatever Has Changed, Before Was Changing

And now that this has been proved, it is evident that everything that is in motion must have been in motion before. For if that which is in motion has traversed the distance KA in the primary time XP, in half the time a thing that is in motion with equal velocity and began its motion at the same time will have traversed half the distance. But if this second thing whose velocity is equal has traversed a certain distance in a certain time, the original thing that is in motion must have traversed the same distance in the same time. Hence that which is in motion must have been in motion before.

Again, if by taking the extreme moment of the time – for it is the moment that defines the time, and time is that which is intermediate between moments – we are enabled to say that motion has taken place in the whole time XP or in fact in any period of it, motion may likewise be said to have taken place in every other such period. But half the time finds an extreme in the point of division. Therefore motion will have taken place in half the time and in fact in any part of it: for as soon as any division is made there is always a time defined by moments. If, then, all time is divisible, and that which is intermediate between moments is time, everything that is changing must have completed an infinite number of changes.

Again, since a thing that changes continuously and has not perished or ceased from its change must either be changing or have changed in any part of the time of its change, and since it cannot be changing in a moment, it follows that it must have changed at every moment in the time: consequently, since the moments are infinite in number, everything that is changing must have completed an infinite number of changes.

And not only must that which is changing have changed, but that which has changed must also previously have been changing, since everything that has changed from something to something has changed in a period of time. For suppose that a thing has changed from A to B in a moment. Now the moment in which it has changed cannot be the same as that in which it is at A (since in that case it would be in A and B at once): for we have shown above that that which has changed, when it has changed, is not in that from which it has changed. If, on the other hand, it is a different moment, there will be a period of time intermediate between the two: for, as we saw, moments are not consecutive. Since, then, it has changed in a period of time, and all time is divisible, in half the time it will have completed another change, in a quarter another, and so on to infinity: consequently when it has changed, it must have previously been changing.

Moreover, the truth of what has been said is more evident in the case of magnitude, because the magnitude over which what is changing changes is continuous. For suppose that a thing has changed from Γ to Δ. Then if $\Gamma\Delta$ is indivisible, two things without parts will be consecutive. But since this is impossible, that which is intermediate between them must be a magnitude and divisible into an infinite number of segments: consequently, before the change is completed, the thing changes to those segments. Everything that has changed, therefore, must previously have been changing: for the same proof also holds good of change with respect to what is not continuous, changes, that is to say, between contraries and between contradictories. In such cases we have only to take the time in which a thing has changed and again apply the same reasoning. So that which has changed must have been changing and that which is changing must have changed, and a process of change is preceded by a completion of change and a completion by a process: and we can never take any stage and say that it is absolutely the first. The reason of this is that no two things without parts can be contiguous, and therefore in change the process of division is infinite, just as lines may be infinitely divided so that one part is continually increasing and the other continually decreasing.

So it is evident also that that which has become must previously have been in process of becoming, and that which is in process of becoming must previously have become, everything (that is) that is divisible and continuous: though it is not always the actual thing that is in process of becoming of which this is true: sometimes it is something else, that is to say, some part of the thing in question, e.g. the foundation-stone of a house. So, too, in the case of that which is perishing and that which has perished: for that which becomes and that which perishes must contain an element of infiniteness as an immediate consequence of the fact that they arc continuous things: and so a thing cannot be in process of becoming without having become or have become without having been in process of becoming. So, too, in the case of perishing and having perished: perishing must be preceded by having perished, and having perished must be preceded by perishing. It is evident, then, that that which has become must previously have been in process of becoming, and that which is in process of becoming must previously have become: for all magnitudes and all periods of time are infinitely divisible.

Consequently no absolutely first stage of change can be represented by any particular part of space or time which the changing thing may occupy.

The Finitude or Infinity of Movement, of Extension, and of the Moved

Now since the motion of everything that is in motion occupies a period of time, and a greater magnitude is traversed in a longer time, it is impossible that a thing should undergo a finite motion in an infinite time, if this is understood to mean not that the same motion or a part of it is continually repeated, but that the whole infinite time is occupied by the whole finite motion. In all cases where a thing is in motion with uniform velocity it is clear that the finite magnitude is traversed in a finite time. For if we take a part of the motion which shall be a measure of the whole, the whole motion is completed in as many equal periods of the time as there are parts of the motion. Consequently, since these parts are finite, both in size individually and in number collectively, the whole time must also be finite: for it will be a multiple of the portion, equal to the time occupied in completing the aforesaid part multiplied by the number of the parts.

But it makes no difference even if the velocity is not uniform. For let us suppose that the line AB represents a finite stretch over which a thing has been moved in the given time, and let ΓΔ be the infinite time. Now if one part of the stretch must have been traversed before another part (this is clear, that in the earlier and in the later part of the time a different part of the stretch has been traversed: for as the time lengthens a different part of the motion will always be completed in it, whether the thing in motion changes with uniform velocity or not: and whether the rate of motion increases or diminishes or remains stationary this is none the less so), let us then take AE a part of the whole stretch of motion AB which shall be a measure of AB. Now this part of the motion occupies a certain period of the infinite time: it cannot itself occupy an infinite time, for we are assuming that that is occupied by the whole AB. And if again I take another part equal to AE, that also must occupy a finite time in consequence of the same assumption. And if I go on taking parts in this way, on the one hand there is no part which will be a measure of the infinite time (for the infinite cannot be composed of finite parts whether equal or unequal, because there must be some unity which will be a measure of things finite in multitude or in magnitude, which, whether they are equal or unequal, are none the less limited in magnitude); while on the other hand the finite stretch of motion AB is a certain multiple of AE: consequently the motion AB must be accomplished in a finite time. Moreover it is the same with coming to rest as with motion. And so it is impossible for one and the same thing to be infinitely in process of becoming or of perishing.

The same reasoning will prove that in a finite time there cannot be an infinite extent of motion or of coming to rest, whether the motion is regular or irregular. For if we take a part which shall be a measure of the whole time, in this part a certain fraction, not the whole, of the magnitude will be traversed, because we assume that the traversing of the whole occupies all the time. Again, in another equal part of the time another part of the magnitude will be traversed: and similarly in each part of the time that we take, whether equal or unequal to the part originally taken. It makes no difference whether the parts are equal or not, if only each is finite: for it is clear that while the time is exhausted by the subtraction of its parts, the infinite magnitude will not be thus exhausted, since the process of subtraction is finite both in respect of the quantity subtracted and of the number of times a subtraction is made. Consequently the infinite magnitude will not be traversed in a finite time: and it makes no difference whether the magnitude is infinite in only one direction or in both: for the same reasoning will hold good.

This having been proved, it is evident that neither can a finite magnitude traverse an infinite magnitude in a finite time, the reason being the same as that given above: in part of the time it will traverse a finite magnitude and in each several part likewise, so that in the whole time it will traverse a finite magnitude.

And since a finite magnitude will not traverse an infinite in a finite time, it is clear that neither will an infinite traverse a finite in a finite time. For if the infinite could traverse the finite, the finite could traverse the infinite; for it makes no difference which of the two is the thing in motion: either case involves the traversing of the infinite by the finite. For when the infinite magnitude A is in motion a part of it, say $\Gamma\Delta$, will occupy the finite B, and then another, and then another, and so on to infinity. Thus the two results will coincide: the infinite will have completed a motion over the finite and the finite will have traversed the infinite: for it would seem to be impossible for the motion of the infinite over the finite to occur in any way other than by the finite traversing the infinite either by locomotion over it or by measuring it. Therefore, since this is impossible, the infinite cannot traverse the finite.

Nor again will the infinite traverse the infinite in a finite time. Otherwise it would also traverse the finite, for the infinite includes the finite. We can further prove this in the same way by taking the time as our starting-point.

Since, then, it is established that in a finite time neither will the finite traverse the infinite, nor the infinite the finite, nor the infinite the infinite, it is evident also that in a finite time there cannot be infinite motion: for what difference does it make whether we take the motion or the magnitude to be infinite? If either of the two is infinite, the other must be so likewise: for all locomotion is in space.

Of Coming to Rest, and of Rest

Since everything to which motion or rest is natural is in motion or at rest in the natural time, place, and manner, that which is coming to a stand, when it is coming to a stand, must be in motion: for if it is not in motion it must be at rest: but that which is at rest cannot be coming to rest. From this it evidently follows that coming to a stand must occupy a period of time: for the motion of that which is in motion occupies a period of time, and that which is coming to a stand has been shown to be in motion: consequently coming to a stand must occupy a period of time.

Again, since the terms 'quicker' and 'slower' are used only of that which occupies a period of time, and the process of coming to a stand may be quicker or slower, the same conclusion follows.

And that which is coming to a stand must be coming to a stand in any part of the primary time in which it is coming to a stand. For if it is coming to a stand in neither of two parts into which the time may be divided, it cannot be coming to a stand in the whole time, with the result that that which is coming to a stand will not be coming to a stand. If on the other hand it is coming to a stand in only one of the two parts of the time, the whole cannot be the primary time in which it is coming to a stand: for it is coming to a stand in the whole time not primarily but in virtue of something distinct from itself, the argument being the same as that which we used above about things in motion.

And just as there is no primary time in which that which is in motion is in motion, so too there is no primary time in which that which is coming to a stand is coming to a stand, there being no primary stage either of being in motion or of coming to a stand. For let AB be the primary time in which a thing is coming to a stand. Now AB cannot be without parts: for there cannot be motion in that which is without parts, because the moving thing would necessarily have been already moved for part of the time of its movement: and that which is coming to a stand has been shown to be in motion. But since AB is therefore divisible, the thing is coming to a stand in every one of the parts of AB: for we have shown above that it is coming to a stand in every one of the parts in which it is primarily coming to a stand. Since, then, that in which primarily a thing is coming to a stand must be a period of time and not something indivisible, and since all time is infinitely divisible, there cannot be anything in which primarily it is coming to a stand.

Nor again can there be a primary time at which the being at rest of that which is at rest occurred: for it cannot have occurred in that which has no parts, because there cannot be motion in that which is indivisible, and that in which rest takes place is the same as that in which motion takes place: for we defined a state of rest to be the state of a thing to which motion is natural but which is not in motion when (that is to say in that in which) motion would be natural to it. Again, our use of the phrase 'being at rest' also implies that the previous state of a thing is still unaltered, not one point only but two at least being thus needed to determine its presence: consequently that in which a thing is at rest cannot be without parts. Since, then, it is divisible, it must be a period of time, and the thing must be at rest in every one of its parts, as may be shown by the same method as that used above in similar demonstrations.

So there can be no primary part of the time: and the reason is that rest and motion are always in a period of time, and a period of time has no primary part any more than a magnitude or in fact anything continuous: for everything continuous is divisible into an infinite number of parts.

A Thing that Is Moved in Any Time Directly Is in No Part of that Time in a Part of the Space Through Which It Moves

And since everything that is in motion is in motion in a period of time and changes from something to something, when its motion is comprised within a particular period of time essentially – that is to say when it fills the whole and not merely a part of the time in question – it is impossible that in that time that which is in motion should be over against some particular thing primarily. For if a thing – itself and each of its parts –occupies the same space for a definite period of time, it is at rest: for it is in just these circumstances that we use the term 'being at rest' – when at one moment after another it can be said with truth that a thing, itself and its parts, occupies the same space. So if this is being at rest it is impossible for that which is changing to be as a whole, at the time when it is primarily changing, over against any particular thing (for the whole period of time is divisible), so that in one part of it after another it will be true to say that the thing, itself and its parts, occupies the same space. If this is not so and the aforesaid proposition is true only at a single moment, then the thing will be over against a particular thing not for any period of time but only at a moment that limits the time. It is true that at any moment it is always over against something stationary: but it is not at rest: for at a moment it is not possible for anything to be either in motion or at rest. So while it is true to say that that which is in motion is at a moment not in motion and is opposite some particular thing,

221

it cannot in a period of time be over against that which is at rest: for that would involve the conclusion that that which is in locomotion is at rest.

Refutation of the Arguments Against the Possibility of Movement

Zeno's reasoning, however, is fallacious, when he says that if everything when it occupies an equal space is at rest, and if that which is in locomotion is always occupying such a space at any moment, the flying arrow is therefore motionless. This is false, for time is not composed of indivisible moments any more than any other magnitude is compose of indivisibles.

Zeno's arguments about motion, which cause so much disquietude to those who try to solve the problems that they present, are four in number. The first asserts the non-existence of motion on the ground that that which is in locomotion must arrive at the half-way stage before it arrives at the goal. This we have discussed above.

The second is the so-called 'Achilles', and it amounts to this, that in a race the quickest runner can never overtake the slowest, since the pursuer must first reach the point whence the pursued started, so that the slower must always hold a lead. This argument is the same in principle as that which depends on bisection, though it differs from it in that the spaces with which we successively have to deal are not divided into halves. The result of the argument is that the slower is not overtaken: but it proceeds along the same lines as the bisection-argument (for in both a division of the space in a certain way leads to the result that the goal is not reached, though the 'Achilles' goes further in that it affirms that even the quickest runner in legendary tradition must fail in his pursuit of the slowest), so that the solution must be the same. And the axiom that that which holds a lead is never overtaken is false: it is not overtaken, it is true, while it holds a lead: but it is overtaken nevertheless if it is granted that it traverses the finite distance prescribed. These then are two of his arguments.

The third is that already given above, to the effect that the flying arrow is at rest, which result follows from the assumption that time is composed of moments: if this assumption is not granted, the conclusion will not follow.

The fourth argument is that concerning the two rows of bodies, each row being composed of an equal number of bodies of equal size, passing each other on a race-course as they proceed with equal velocity in opposite directions, the one row originally occupying the space between the goal and the middle point of the course and the other that between the middle point and the starting-post. This, he thinks, involves the conclusion that half a given time is equal to double that time. The fallacy of the reasoning lies in the assumption that a body occupies an equal time in passing with equal velocity a body that is in motion and a body of equal size that is at rest; which is false. For instance (so runs the argument), let A, A...be the stationary bodies of equal size, B, B...the bodies, equal in number and in size to A, A..., originally occupying the half of the course from the starting-post to the middle of the As, and Γ, Γ...those originally occupying the other half from the goal to the middle of the As, equal in number, size, and velocity to B, B...Then three consequences follow:

First, as the Bs and the Γs pass one another, the first B reaches the last Γ at the same moment as the first Γ reaches the last B. Secondly, at this moment the first Γ has passed all the As, whereas the first B has passed only half the As, and has consequently occupied only half the time occupied by the first Γ, since each of the two occupies an equal time in passing each A. Thirdly, at the same moment all the Bs have passed all the Γs: for the first Γ and the first B will simultaneously reach the opposite ends of the course, since (so says Zeno) the time occupied by the first Γ in passing each of the Bs is equal to that occupied by it in passing each of the As, because an equal time is occupied by both the first B and the first Γ in passing all the As. This is the argument, but it presupposed the aforesaid fallacious assumption.

Nor in reference to contradictory change shall we find anything unanswerable in the argument that if a thing is changing from not-white, say, to white, and is in neither condition, then it will be neither white nor not-white: for the fact that it is not *wholly* in either condition will not preclude us from calling it white or not-white. We call a thing white or not-white not necessarily because it is wholly either one or the other, but because most of its parts or the most essential parts of it are so: not being in a certain condition is different from not being wholly in that condition. So, too, in the case of being and not-being and all other conditions which stand in a contradictory relation: while the changing thing must of necessity be in one of the two opposites, it is never wholly in either.

Again, in the case of circles and spheres and everything whose motion is confined within the space that it occupies, it is not true to say that the motion can be nothing but rest, on the ground that such things in motion, themselves and their parts, will occupy the same position for a period of time, and that therefore they will be at once at rest and in motion. For in the first place the parts do not occupy the same position for any period of time: and in the second place the whole also is always changing to a different position: for if we take the orbit as described from a point A on a circumference, it will not be the same as the orbit as described from B or Γ or any other point on the same circumference except in an accidental sense, the sense that is to say in which a musical man is the same as a man. Thus one orbit is always changing into another, and the thing will never be at rest. And it is the same with the sphere and everything else whose motion is confined within the space that it occupies.

That Which Has Not Parts Cannot Move

Our next point is that that which is without parts cannot be in motion except accidentally: i.e. it can be in motion only in so far as the body or the magnitude is in motion and the partless is in motion by inclusion therein, just as that which is in a boat may be in motion in consequence of the locomotion of the boat, or a part may be in motion in virtue of the motion of the whole. (It must be remembered, however, that by 'that which is without parts' I mean that which is quantitatively indivisible (and that the case of the motion of a part is not exactly parallel): for parts have motions belonging essentially and severally to themselves distinct from the motion of the whole. The distinction may be seen most clearly in the case of a revolving sphere, in which the velocities of the parts near the centre and of those on the surface are different from one another and from that of the whole; this implies that there is not one motion but many.) As we have said, then, that which is without parts can be in motion in the sense in which a man sitting in a boat is in motion when the boat is travelling, but it cannot be in motion of itself. For suppose that it is changing from AB to BΓ – either from one magnitude to another, or from one form to another, or from some state to its contradictory – and let Δ be the primary time in which it undergoes the change. Then in the time in which it is changing it must be either in AB or in B Γ or partly in one and partly in the other: for this, as we saw, is true of everything that is changing. Now it cannot be partly in each of the two: for then it would be divisible into parts. Nor again can it be in BΓ: for then it will have completed the change, whereas the assumption is that the change is in process. It remains, then, that in the time in which it is changing, it is in AB. That being so, it will be at rest: for, as we saw, to be in the same condition for a period of time is to be at rest. So it is not possible for that which has no parts to be in motion or to change in any way: for only one condition could have made it possible for it to have motion, viz. that time should be composed of moments, in which case at any moment it would have completed a motion or a change, so that it would never be in motion, but would always have been in motion. But this we have already shown above to be impossible: time is not composed of moments, just as a line is not composed of points, and motion is not composed of starts: for this theory simply makes motion consist of indivisibles in exactly the same way as time is made to consist of moments or a length of points.

Again, it may be shown in the following way that there can be no motion of a point or of any other indivisible. That which is in motion can never traverse a space greater than itself without first traversing a space equal to or less than itself. That being so, it is evident that the point also must first traverse a space equal to or less than itself. But since it is indivisible, there can be no space less than itself for it to traverse first: so it will have to traverse a distance equal to itself. Thus the line will be composed of points, for the point, as it continually traverses a distance equal to itself, will be a measure of the whole line. But since this is impossible, it is likewise impossible for the indivisible to be in motion.

Again, since motion is always in a period of time and never in a moment, and all time is divisible, for everything that is in motion there must be a time less than that in which it traverses a distance as great as itself. For that in which it is in motion will be a time, because all motion is in a period of time; and all time has been shown above to be divisible. Therefore, if a point is in motion, there must be a time less than that in which it has itself traversed any distance. But this is impossible, for in less time it must traverse less distance, and thus the indivisible will be divisible into something less than itself, just as the time is so divisible: the fact being that the only condition under which that which is without parts and indivisible could be in motion would have been the possibility of the infinitely small being in motion in a moment: for in the two questions – that of motion in a moment and that of motion of something indivisible –the same principle is involved.

Can Change Be Infinite?

Our next point is that no process of change is infinite: for every change, whether between contradictories or between contraries, is a change from something to something. Thus in contradictory changes the positive or the negative, as the case may be, is the limit, e.g. being is the limit of coming to be and not-being is the limit of ceasing to be: and in contrary changes the particular contraries are the limits, since these are the extreme points of any such process of change, and consequently of every process of alteration: for alteration is always dependent upon some contraries. Similarly contraries are the extreme points of processes of increase and decrease: the limit of increase is to be found in the complete magnitude proper to the peculiar nature of the thing that is increasing, while the limit of decrease is the complete loss of such magnitude. Locomotion, it is true, we cannot show to be finite in this way, since it is not always between contraries. But since that which cannot be cut (in the sense that it is inconceivable that it should be cut, the term 'cannot' being used in several senses) – since it is inconceivable that that which in this sense cannot be cut should be in process of being cut, and generally that that which cannot come to be should be in process of coming to be, it follows that it is inconceivable that that which cannot complete a change should be in process of changing to that to which it cannot complete a change. If, then, it is to be assumed that that which is in locomotion is in process of changing, it must be capable of completing the change. Consequently its motion is not infinite, and it will not be in locomotion over an infinite distance, for it cannot traverse such a distance.

It is evident, then, that a process of change cannot be infinite in the sense that it is not defined by limits. But it remains to be considered whether it is possible in the sense that one and the same process of change may be infinite in respect of the time which it occupies. If it is not one process, it would seem that there is nothing to prevent its being infinite in this sense; e.g. if a process of locomotion be succeeded by a process of alteration and that by a process of increase and that again by a process of coming to be: in this way there may be motion for ever so far as the time is concerned, but it will not be one motion, because all these motions do not compose one. If it is to be one process, no motion can be infinite in respect of the time that it occupies, with the single exception of rotatory locomotion.

Book VII

Whatever Is Moved Is Moved by Something

Everything that is in motion must be moved by something. For if it has not the source of its motion in itself it is evident that it is moved by something other than itself, for there must be something else that moves it. If on the other hand it has the source of its motion in itself, let AB be taken to represent that which is in motion essentially of itself and not in virtue of the fact that something belonging to it is in motion. Now in the first place to assume that AB, because it is in motion as a whole and is not moved by anything external to itself, is therefore moved by itself – this is just as if, supposing that KΛ is moving ΛM and is also itself in motion, we were to deny that KM is moved by anything on the ground that it is not evident which is the part that is moving it and which the part that is moved. In the second place that which is in motion without being moved by anything does not necessarily cease from its motion because something else is at rest, but a thing must be moved by something if the fact of something else having ceased from its motion causes it to be at rest. Thus, if this is accepted, everything that is in motion must be moved by something. For AB, which has been taken to represent that which is in motion, must be divisible, since everything that is in motion is divisible. Let it be divided, then, at Γ. Now if ΓB is not in motion, then AB will not be in motion: for if it is, it is clear that AΓ would be in motion while BΓ is at rest, and thus AB cannot be in motion essentially and primarily. But *ex hypothesi* AB is in motion essentially and primarily. Therefore if ΓB is not in motion AB will be at rest. But we have agreed that that which is at rest if something else is not in motion must be moved by something. Consequently, everything that is in motion must be moved by something: for that which is in motion will always be divisible, and if a part of it is not in motion the whole must be at rest.

There Is a First Movent Which Is Not Moved by Anything Else

Since everything that is in motion must be moved by something, let us take the case in which a thing is in locomotion and is moved by something that is itself in motion, and that again is moved by something else that is in motion, and that by something else, and so on continually: then the series cannot go on to infinity, but there must be some first movent. For let us suppose that this is not so and take the series to be infinite. Let

A then be moved by B, B by Γ, Γ by Δ, and so on, each member
of the series being moved by that which comes next to it. Then
since *ex hypothesi* the movent while causing motion is also itself
in motion, and the motion of the moved and the motion of the
movent must proceed simultaneously (for the movent is causing
motion and the moved is being moved simultaneously) it is
evident that the respective motions of A, B, Γ, and each of the
other moved movents are simultaneous. Let us take the motion
of each separately and let E be the motion of A, Z of B, and
H and Θ respectively the motions of Γ and Δ: for though they
are all moved severally one by another, yet we may still take
the motion of each as numerically one, since every motion is
from something to something and is not infinite in respect of its
extreme points. By a motion that is numerically one I mean a
motion that proceeds from something numerically one and the
same to something numerically one and the same in a period
of time numerically one and the same: for a motion may be the
same generically, specifically, or numerically: it is generically
the same if it belongs to the same category, e.g. substance or
quality: it is specifically the same if it proceeds from something
specifically the same to something specifically the same, e.g.
from white to black or from good to bad, which is not of a kind
specifically distinct: it is numerically the same if it proceeds
from something numerically one to something numerically one
in the same period of time, e.g. from a particular white to a
particular black, or from a particular place to a particular place,
in a particular period of time: for if the period of time were not
one and the same, the motion would no longer be numerically

one though it would still be specifically one. We have dealt with this question above. Now let us further take the time in which A has completed its motion, and let it be represented by K. Then since the motion of A is finite the time will also be finite. But since the movents and the things moved are infinite, the motion EZHΘ, i.e. the motion that is composed of all the individual motions, must be infinite. For the motions of A, B, and the others may be equal, or the motions of the others may be greater: but assuming what is conceivable, we find that whether they are equal or some are greater, in both cases the whole motion is infinite. And since the motion of A and that of each of the others are simultaneous, the whole motion must occupy the same time as the motion of A: but the time occupied by the motion of A is finite: consequently the motion will be infinite in a finite time, which is impossible.

It might be thought that what we set out to prove has thus been shown, but our argument so far does not prove it, because it does not yet prove that anything impossible results from the contrary supposition: for in a finite time there may be an infinite motion, though not of one thing, but of many: and in the case that we are considering this is so: for each thing accomplishes its own motion, and there is no impossibility in many things being in motion simultaneously. But if (as we see to be universally the case) that which primarily is moved locally and corporeally must be either in contact with or continuous with that which moves it, the things moved and the movents must be continuous or in contact with one another, so that together they all form a single unity: whether this unity is finite or infinite makes no difference to our present argument; for in any case since the things in motion are infinite in number the whole motion will be infinite, if, as is theoretically possible, each motion is either equal to or greater than that which follows it in the series: for we shall take as actual that which is theoretically possible. If, then, A, B, Γ, Δ form an infinite magnitude that passes through the motion EZHΘ in the finite time K, this involves the conclusion that an infinite motion is passed through in a finite time: and whether the magnitude in question is finite or infinite this is in either case impossible. Therefore the series must come to an end, and there must be a first movent and a first moved: for the fact that this impossibility results only from the assumption of a particular case is immaterial, since the case assumed is theoretically possible, and the assumption of a theoretically possible case ought not to give rise to any impossible result.

The Movent and the Moved
Are Together

That which is the first movent of a thing – in the sense that it supplies not 'that for the sake of which' but the source of the motion – is always together with that which is moved by it (by 'together' I mean that there is nothing intermediate between them). This is universally true wherever one thing is moved by another. And since there are three kinds of motion, local, qualitative, and quantitative, there must also be three kinds of movent, that which causes locomotion, that which causes alteration, and that which causes increase or decrease.

Let us begin with locomotion, for this is the primary motion. Everything that is in locomotion is moved either by itself or by something else. In the case of things that are moved by themselves it is evident that the moved and the movent are together: for they contain within themselves their first movent, so that there is nothing in between. The motion of things that are moved by something else must proceed in one of four ways: for there are four kinds of locomotion caused by something other than that which is in motion, viz. pulling, pushing, carrying, and twirling. All forms of locomotion are reducible to these. Thus pushing on is a form of pushing in which that which is causing motion away from itself follows up that which it pushes and continues to push it: pushing off occurs when the movent does not follow up the thing that it has moved: throwing when the movent causes a motion away from itself more violent than the natural locomotion of the thing moved, which continues its course so long as it is controlled by the motion imparted to it. Again, pushing apart and pushing together are forms respectively of pushing off and pulling: pushing apart is pushing off, which may be a motion either away from the pusher or away from something else, while pushing together is pulling, which may be a motion towards something else as well as towards the puller. We may similarly classify all the varieties of these last two, e.g. packing and combing: the former is a form of pushing together, the latter a form of pushing apart. The same is true of the other processes of combination and separation (they will all be found to be forms of pushing apart or of pushing together), except such as are involved in the processes of becoming and perishing. (At the same time it is evident that there is no other kind of motion but combination and separation: for they may all be apportioned to one or other of those already mentioned.) Again, inhaling is a form of pulling, exhaling a form of pushing: and the same is true of spitting and of all other motions that proceed through the body, whether secretive or assimilative, the assimilative being forms of pulling, the secretive of pushing off. All other kinds of locomotion must be similarly reduced, for they all fall under one or other of our four heads. And again, of these four, carrying and twirling are reducible to pulling and pushing. For carrying always follows one of the other three methods, for that which is carried is in motion accidentally, because it is in or upon something that is in motion, and that which carries it is in doing so being either pulled or pushed or twirled; thus carrying belongs to all the other three kinds of motion in common. And twirling is a compound of pulling and pushing, for that which is twirling a thing must be pulling one part of the thing and pushing another part, since it impels one part away from itself and another part towards itself. If, therefore, it can be shown that that which is pushing and that which is pulling are adjacent respectively to that which is being pushed and that which is being pulled, it will be evident that in all locomotion there is nothing intermediate between moved and movent. But the former fact is clear even from the definitions of pushing and pulling, for pushing is motion to something else from oneself or from something else, and pulling is motion from something else to oneself or to something else, when the motion of that which is pulling is quicker than the motion that would separate from one another the two things that are continuous: for it is this that causes one thing to be pulled on along with the other. (It might indeed be thought that there is a form of pulling that arises in another way: that wood, e.g. pulls fire in a manner different from that described above. But it makes no difference whether that which pulls is in motion or is stationary when it is pulling: in the latter case it pulls to the place where it is, while in the former it pulls to the place where it was.) Now it is impossible to move anything either from oneself to something else or from something else to oneself without being in contact with it: it is evident, therefore, that in all locomotion there is nothing intermediate between moved and movent.

Nor again is there anything intermediate between that which undergoes and that which causes alteration: this can be proved by induction: for in every case we find that the respective extremities of that which causes and that which undergoes alteration are adjacent. For our assumption is that things that are undergoing alteration are altered in virtue of their being affected in respect of their so-called affective qualities, since that which is of a certain quality is altered in so far as it is sensible, and the characteristics in which bodies differ from one another are sensible characteristics: for every body differs from another in possessing a greater or lesser number of sensible characteristics or in possessing the same sensible characteristics in a greater or lesser degree. But the alteration of that which undergoes alteration is also caused by

the above-mentioned characteristics, which are affections of
some particular underlying quality. Thus we say that a thing is
altered by becoming hot or sweet or thick or dry or white: and
we make these assertions alike of what is inanimate and of what
is animate, and further, where animate things are in question,
we make them both of the parts that have no power of sense-
perception and of the senses themselves. For in a way even the
senses undergo alteration, since the active sense is a motion
through the body in the course of which the sense is affected
in a certain way. We see, then, that the animate is capable of
every kind of alteration of which the inanimate is capable: but
the inanimate is not capable of every kind of alteration of which
the animate is capable, since it is not capable of alteration in
respect of the senses: moreover the inanimate is unconscious of
being affected by alteration, whereas the animate is conscious
of it, though there is nothing to prevent the animate also being
unconscious of it when the process of the alteration does not
concern the senses. Since, then, the alteration of that which
undergoes alteration is caused by sensible things, in every case
of such alteration it is evident that the respective extremities
of that which causes and that which undergoes alteration are
adjacent. Thus the air is continuous with that which causes the
alteration, and the body that undergoes alteration is continuous
with the air. Again, the colour is continuous with the light and
the light with the sight. And the same is true of hearing and
smelling: for the primary movent in respect to the moved is
the air. Similarly, in the case of tasting, the flavour is adjacent
to the sense of taste. And it is just the same in the case of

things that are inanimate and incapable of sense-perception. Thus there can be nothing intermediate between that which undergoes and that which causes alteration.

Nor, again, can there be anything intermediate between that which suffers and that which causes increase: for the part of the latter that starts the increase does so by becoming attached in such a way to the former that the whole becomes one. Again, the decrease of that which suffers decrease is caused by a part of the thing becoming detached. So that which causes increase and that which causes decrease must be continuous with that which suffers increase and that which suffers decrease respectively: and if two things are continuous with one another there can be nothing intermediate between them.

It is evident, therefore, that between the extremities of the moved and the movent that are respectively first and last in reference to the moved there is nothing intermediate.

All Alteration Pertains to Sensible Qualities

Everything, we say, that undergoes alteration is altered by sensible causes, and there is alteration only in things that are said to be essentially affected by sensible things. The truth of this is to be seen from the following considerations. Of all other things it would be most natural to suppose that there is alteration in figures and shapes, and in acquired states and in the processes of acquiring and losing these: but as a matter of fact in neither of these two classes of things is there alteration.

In the first place, when a particular formation of a thing is completed, we do not call it by the name of its material: e.g. we do not call the statue 'bronze' or the pyramid 'wax' or the bed 'wood', but we use a derived expression and call them 'of bronze', 'waxen,' and 'wooden' respectively. But when a thing has been affected and altered in any way we still call it by the original name: thus we speak of the bronze or the wax being dry or fluid or hard or hot.

And not only so: we also speak of the particular fluid or hot substance as being bronze, giving the material the same name as that which we use to describe the affection.

Since, therefore, having regard to the figure or shape of a thing we no longer call that which has become of a certain figure by the name of the material that exhibits the figure, whereas having regard to a thing's affections or alterations we still call it by the name of its material, it is evident that becomings of the former kind cannot be alterations.

Moreover it would seem absurd even to speak in *this* way, to speak, that is to say, of a man or house or anything else that has come into existence as having been altered. Though it may be true that every such becoming is necessarily the result of something's being altered, the result, e.g. of the material's being condensed or rarefied or heated or cooled, nevertheless it is not the things that are coming into existence that are altered, and their becoming is not an alteration.

Again, acquired states, whether of the body or of the soul, are not alterations. For some are excellences and others are defects, and neither excellence nor defect is an alteration:

excellence is a perfection (for when anything acquires its proper excellence we call it perfect, since it is then if ever that we have a thing in its natural state: e.g. we have a perfect circle when we have one as good as possible), while defect is a perishing of or departure from this condition. So just as when speaking of a house we do not call its arrival at perfection an alteration (for it would be absurd to suppose that the coping or the tiling is an alteration or that in receiving its coping or its tiling a house is altered and not perfected), the same also holds good in the case of excellences and defects and of the persons or things that possess or acquire them: for excellences are perfections of a thing's nature and defects are departures from it: consequently they are not alterations.

Further, we say that all excellences depend upon particular relations. Thus bodily excellences such as health and a good state of body we regard as consisting in a blending of hot and cold elements within the body in due proportion, in relation either to one another or to the surrounding atmosphere: and in like manner we regard beauty, strength, and all the other bodily excellences and defects. Each of them exists in virtue of a particular relation and puts that which possesses it in a good or bad condition with regard to its proper affections, where by 'proper' affections I mean those influences that from the natural constitution of a thing tend to promote or destroy its existence. Since, then, relatives are neither themselves alterations nor the subjects of alteration or of becoming or in fact of any change whatever, it is evident that neither states nor the processes of losing and acquiring states are alterations, though it may be

241

true that their becoming or perishing is necessarily, like the becoming or perishing of a specific character or form, the result of the alteration of certain other things, e.g. hot and cold or dry and wet elements or the elements, whatever they may be, on which the states primarily depend. For each several bodily defect or excellence involves a relation with those things from which the possessor of the defect or excellence is naturally subject to alteration: thus excellence disposes its possessor to be unaffected by these influences or to be affected by those of them that ought to be admitted, while defect disposes its possessor to be affected by them or to be unaffected by those of them that ought to be admitted.

And the case is similar in regard to the states of the soul, all of which (like those of body) exist in virtue of particular relations, the excellences being perfections of nature and the defects departures from it: moreover, excellence puts its possessor in good condition, while defect puts its possessor in a bad condition, to meet his proper affections. Consequently these cannot any more than the bodily states be alterations, nor can the processes of losing and acquiring them be so, though their becoming is necessarily the result of an alteration of the sensitive part of the soul, and this is altered by sensible objects: for all moral excellence is concerned with bodily pleasures and pains, which again depend either upon acting or upon remembering or upon anticipating. Now those that depend upon action are determined by sense-perception, i.e. they are stimulated by something sensible: and those that depend upon memory or anticipation are likewise to be traced to sense-perception, for in these cases pleasure is felt either in remembering what one has experienced or in anticipating what one is going to experience. Thus all pleasure of this kind must be produced by sensible things: and since the presence in anyone of moral defect or excellence involves the presence in him of pleasure or pain (with which moral excellence and defect are always concerned), and these pleasures and pains are alterations of the sensitive part, it is evident that the loss and acquisition of these states no less than the loss and acquisition of the states of the body must be the result of the alteration of something else. Consequently, though their becoming is accompanied by an alteration, they are not themselves alterations.

Again, the states of the intellectual part of the soul are not alterations, nor is there any becoming of them. In the first place it is much more true of the possession of knowledge that it depends upon a particular relation. And further, it is evident that there is no becoming of these states. For that which is potentially possessed of knowledge becomes actually possessed of it not by being set in motion at all itself but by reason of the presence of something else: i.e. it is when it meets with the particular object that it knows in a manner the particular through its knowledge of the universal. (Again, there is no becoming of the actual use and activity of these states, unless it is thought that there is a becoming of vision and touching and that the activity in question is similar to these.) And the original acquisition of knowledge is not a becoming or an alteration: for the terms 'knowing' and 'understanding' imply that the intellect has reached a state of rest and come to a standstill, and there is no becoming that leads to a state of rest, since, as we have said above, no change at all can have a becoming. Moreover, just as to say, when anyone has passed from a state of intoxication or sleep or disease to the contrary state, that he has become possessed of knowledge again is incorrect in spite of the fact that he was previously incapable of using his knowledge, so, too, when anyone originally acquires the state, it is incorrect to say that he becomes possessed of knowledge: for the possession of understanding and knowledge is produced by the soul's settling down out of the restlessness natural to it. Hence, too, in learning and in forming judgements on matters relating to their sense-perceptions children are inferior to adults owing to the great amount of restlessness and motion in their souls. Nature itself causes the soul to settle down and come to a state of rest for the performance of some of its functions, while for the performance of others other things do so: but in either case the result is brought about through the alteration of something in the body, as we see in the case of the use and activity of the intellect arising from a man's becoming sober or being awakened. It is evident, then, from the preceding argument that alteration and being altered occur in sensible things and in the sensitive part of the soul and, except accidentally, in nothing else.

Comparison of Movements

A difficulty may be raised as to whether every motion is commensurable with every other or not. Now if they are all commensurable and if two things to have the same velocity must accomplish an equal motion in an equal time, then we may have a circumference equal to a straight line, or, of course, the one may be greater or less than the other. Further, if one thing alters and another accomplishes a locomotion in an equal time, we may have an alteration and a locomotion equal to one another: thus an affection will be equal to a length, which is impossible. But is it not only when an equal motion is accomplished by two things in an equal time that the velocities of the two are equal? Now an affection cannot be equal to a length. Therefore there cannot be an alteration equal to or less than a locomotion: and consequently it is not the case that every motion is commensurable with every other.

But how will our conclusion work out in the case of the circle and the straight line? It would be absurd to suppose that the motion of one thing in a circle and of another in a straight line cannot be similar, but that the one must inevitably move more quickly or more slowly than the other, just as if the course of one were downhill and of the other uphill. Moreover it does not as a matter of fact make any difference to the argument to say that the one motion must inevitably be quicker or slower than the other: for then the circumference can be greater or less than the straight line; and if so it is possible for the two to be equal. For if in the time A the quicker (B) passes over , the distance B' and the slower (Γ) passes over the distance Γ', B' will be greater than Γ': for this is what we took 'quicker' to mean: and so quicker motion also implies that one thing traverses an equal distance in less time than another: consequently there will be a part of A in

which Β will pass over a part of the circle equal to Γ', while Γ
will occupy the whole of A in passing over Γ'. None the less, if
the two motions are commensurable, we are confronted with the
consequence stated above, viz. that there may be a straight line
equal to a circle. But these are not commensurable: and so the
corresponding motions are not commensurable either.

But may we say that things are always commensurable if the
same terms are applied to them without equivocation? e.g. a pen,
a wine, and the highest note in a scale are not commensurable:
we cannot say whether any one of them is sharper than any
other: and why is this? they are incommensurable because it is
only equivocally that the same term 'sharp' is applied to them:
whereas the highest note in a scale is commensurable with the
leadingnote, because the term 'sharp' has the same meaning as
applied to both. Can it be, then, that the term 'quick' has not
the same meaning as applied to straight motion and to circular
motion respectively? If so, far less will it have the same meaning
as applied to alteration and to locomotion.

Or shall we in the first place deny that things are always
commensurable if the same terms are applied to them without
equivocation? For the term 'much' has the same meaning
whether applied to water or to air, yet water and air are not
commensurable in respect of it: or, if this illustration is not
considered satisfactory, 'double' at any rate would seem to
have the same meaning as applied to each (denoting in each
case the proportion of two to one), yet water and air are not
commensurable in respect of it. But here again may we not take
up the same position and say that the term 'much' is equivocal?

In fact there are some terms of which even the definitions are equivocal; e.g. if 'much' were defined as 'so much and more', 'so much' would mean something different in different cases: 'equal' is similarly equivocal; and 'one' again is perhaps inevitably an equivocal term; and if 'one' is equivocal, so is 'two'. Otherwise why is it that some things are commensurable while others are not, if the nature of the attribute in the two cases is really one and the same?

Can it be that the incommensurability of two things in respect of any attribute is due to a difference in that which is primarily capable of carrying the attribute? Thus horse and dog are so commensurable that we may say which is the whiter, since that which primarily contains the whiteness is the same in both, viz. the surface: and similarly they are commensurable in respect of size. But water and speech are not commensurable in respect of clearness, since that which primarily contains the attribute is different in the two cases. It would seem, however, that we must reject this solution, since clearly we could thus make all equivocal attributes univocal and say merely that that which contains each of them is different in different cases: thus 'equality', 'sweetness', and 'whiteness' will severally always be the same, though that which contains them is different in different cases. Moreover, it is not any casual thing that is capable of carrying any attribute: each single attribute can be carried primarily only by one single thing.

Must we then say that, if two things are to be commensurable in respect of any attribute, not only must the attribute in question be applicable to both without equivocation, but there must also be no specific differences either in the attribute itself or in that which contains the attribute – that these, I mean, must not be divisible in the way in which colour is divided into kinds? Thus in this respect one thing will not be commensurable with another, i.e. we cannot say that one is more coloured than the other where only colour in general and not any particular colour is meant; but they are commensurable in respect of whiteness.

Similarly in the case of motion: two things are of the same velocity if they occupy an equal time in accomplishing a certain equal amount of motion. Suppose, then, that in a certain time an alteration is undergone by one half of a body's length and a locomotion is accomplished by the other half: can we say that in this case the alteration is equal to the locomotion and of the same velocity? That would be absurd, and the reason is that there are different species of motion. And if in consequence of this we must say that two things are of equal velocity if they accomplish locomotion over an equal distance in an equal time, we have to admit the equality of a straight line and a circumference. What, then, is the reason of this? Is it that locomotion is a genus or that line is a genus? (We may leave the time out of account, since that is one and the same.) If the lines are specifically different, the locomotions also differ specifically from one another: for locomotion is specifically differentiated according to the specific differentiation of that over which it takes place. (It is also similarly differentiated, it would seem, accordingly as the instrument of the locomotion is different: thus if feet are the instrument, it is walking, if wings it is flying; but perhaps we should rather say that this is not so, and that in this case the differences in the locomotion are merely differences of posture in that which is in motion.)We may say, therefore, that things are of equal velocity if in an equal time they traverse the same magnitude: and when I call it 'the same' I mean that it contains no specific difference and therefore no difference in the motion that takes place over it. So we have now to consider how motion is differentiated: and this discussion serves to show that the genus is not a unity but contains a plurality latent in it and distinct from it, and that in the case of equivocal terms sometimes the different senses in which they are used are far removed from one another, while sometimes there is a certain likeness between them, and sometimes again they are nearly related either generically or analogically, with the result that they seem not to be equivocal though they really are.

When, then, is there a difference of species? Is an attribute specifically different if the subject is different while the attribute is the same, or must the attribute itself be different as well? And how are we to define the limits of a species? What will enable us to decide that particular instances of whiteness or sweetness are the same or different? Is it enough that it appears different in one subject from what it appears in another? Or must there be no sameness at all? And further, where alteration is in question, how is one alteration to be of equal velocity with another? One person may be cured quickly and another slowly, and cures may also be simultaneous: so that, recovery of health being an alteration, we have here alterations of equal velocity, since each alteration occupies an equal time. But what alteration? We cannot here speak of an 'equal' alteration: what corresponds in the category of quality to equality in the category of quantity is 'likeness'.

However, let us say that there is equal velocity where *the same* change is accomplished in an equal time. Are we, then, to finds the commensurability in the subject of the affection or in the affection itself? In the case that we have just been considering it is the fact that health is one and the same that enables us to arrive at the conclusion that the one alteration is neither more nor less than the other, but that both are alike. If on the other hand the affection is different in the two cases, e.g. when the alterations take the form of becoming white and becoming healthy respectively, here there is no sameness or equality or likeness inasmuch as the difference in the affections at once makes the alterations specifically different, and there is no unity of alteration any more than there would be unity of locomotion under like conditions. So we must find out how many species there are of alteration and of locomotion respectively. Now if the things that are in motion – that is to say, the things to which the motions belong essentially and not accidentally – differ specifically, then their respective motions will also differ specifically: if on the other hand they differ generically or numerically, the motions also will differ generically or numerically as the case may be. But there still remains the question whether, supposing that two alterations are of equal velocity, we ought to look for this equality in the sameness (or likeness) of the affections, or in the things altered, to see e.g. whether a certain quantity of each has become white. Or ought we not rather to look for it in both? That is to say, the alterations are the same or different according as the affections are the same or different, while they are equal or unequal according as the things altered are equal or unequal.

And now we must consider the same question in the case of becoming and perishing: how is one becoming of equal velocity with another? They are of equal velocity if in an equal time there are produced two things that are the same and specifically inseparable, e.g. two men (not merely generically inseparable as e.g. two animals). Similarly one is quicker than the other if in an equal time the product is different in the two cases. I state it thus because we have no pair of terms that will convey this 'difference' in the way in which unlikeness is conveyed. If we adopt the theory that it is number that constitutes being, we may indeed speak of a 'greater number' and a 'lesser number' within the same species, but there is no common term that will include both relations, nor are there terms to express each of them separately in the same way as we indicate a higher degree or preponderance of an affection by 'more', of a quantity by 'greater'.

Proportion of Movements

Now since wherever there is a movent, its motion always acts upon something, is always in something, and always extends to something (by 'is always in something' I mean that it occupies a time: and by 'extends to something' I mean that it involves the traversing of a certain amount of distance: for at any moment when a thing is causing motion, it also has caused motion, so that there must always be a certain amount of distance that has been traversed and a certain amount of time that has been occupied).

If, then, A the movent have moved B a distance Γ in a time Δ, then in the same time the same force A will move ½B twice the distance Γ, and in ½Δ it will move ½B the whole distance Γ: for thus the rules of proportion will be observed. Again if a given force move a given weight a certain distance in a certain time and half the distance in half the time, half the motive power will move half the weight the same distance in the same time. Let E represent half the motive power A and Z half the weight B: then the ratio between the motive power and the weight in the one case is similar and proportionate to the ratio in the other, so that each force will cause the same distance to be traversed in the same time.

But if E move Z a distance Γ in a time Δ, it does not necessarily follow that E can move twice Z half the distance Γ in the same time. If, then, A move B a distance Γ in a time Δ, it does not follow that E, being half of A, will in the time Δ or in any fraction of it cause B to traverse a part of Γ the ratio between which and the whole of Γ is proportionate to that between A and E (whatever fraction of A E may be): in fact it might well be that it will cause no motion at all; for it does not follow that, if a given motive power causes a certain amount of motion, half that power will cause motion either of any particular amount or in any length of time: otherwise one man might move a ship, since both the motive power of the shiphaulers and the distance that they all cause the ship to traverse are divisible into as many parts as there are men. Hence Zeno's reasoning is false when he argues that there is no part of the millet that does not make a sound: for there is no reason why any such part should not in any length of time fail to move the air that the whole bushel moves in falling.

In fact it does not of itself move even such a quantity of the air as it would move if this part were by itself: for no part even exists otherwise than potentially. If on the other hand we have two forces each of which separately moves one of two weights a given distance in a given time, then the forces in combination will move the combined weights an equal distance in an equal time: for in this case the rules of proportion apply.

Then does this hold good of alteration and of increase also? Surely it does, for in any given case we have a definite thing that causes increase and a definite thing that suffers increase, and the one causes and the other suffers a certain amount of increase in a certain amount of time. Similarly we have a definite thing that causes alteration and a definite thing that undergoes alteration, and a certain amount, or rather degree, of alteration is completed in a certain amount of time: thus in twice as much time twice as much alteration will be completed and conversely twice as much alteration will occupy twice as much time: and the alteration of half of its object will occupy half as much time and in half as much time half of the object will be altered: or again, in the same amount of time it will be altered twice as much.

On the other hand if that which causes alteration or increase causes a certain amount of increase or alteration respectively in a certain amount of time, it does not necessarily follow that half the force will occupy twice the time in altering or increasing the object, or that in twice the time the alteration or increase will be completed by it: it may happen that there will be no alteration or increase at all, the case being the same as with the weight.

Biographies

Aristotle

The Physics

The Ancient Greek philosopher Aristotle (384–322 BCE) has had a monumental influence on the development of countless scientific and cultural fields of study, from physics to zoology, and from aesthetics to linguistics. His writings have served as cornerstones in the foundation of Western society, and his development of the Aristotelian tradition, based in deductive logic and analysis, has for millennia shaped the methods of the scientific pursuit of knowledge. Aristotle's founding of the Peripatetic school of philosophy at the Lyceum in Athens helped develop and disseminate this tradition for the wider world.

R.P. Hardie and R.K. Gaye

R.P. Hardie (1864–1942) was a translator of Ancient Greek and a scholar of classical antiquity. Working with his colleague R.K. Gaye (1877–1909), who was also a translator and classical scholar, together they translated numerous texts by

Ancient Greek authors. One of their best-known works is their translation of *The Physics* by Aristotle, a collection of treatises on foundational philosophical principles concerning natural and moving objects. Hardie also co-edited Robert Adamson's *The Development of Greek Philosophy* (1908). R.K. Gaye was the author of *The Platonic Conception of Immortality and Its Connexion with the Theory of Ideas* (1904).

James Lees
The New Introduction to Aristotle
Dr James Lees is a physicist, author, and science communicator who studied at the University of York and Stanford. He works as a researcher and can be found giving talks and workshops across the UK at various educational festivals on a wide variety of scientific topics. He also has several other books to his name, including *Physics in 50 Milestone Moments* and *Who Knew? Physics.*

Marika Taylor
Series Foreword
Professor Marika Taylor is a Professor of Theoretical Physics and Head of School within Mathematical Sciences at the University of Southampton. Her research interests include all aspects of string theory, gravitational physics and quantum field theory. In recent years much of her work has been focused on holographic dualities and their implications. Marika's research has featured in such publications as *Physical Review, Journal of High Energy Physics* and *General Relativity and Gravitation* among others.

With introductions and a new foreword.

A new series of accessible books bringing
together ancient, medieval and modern texts in a
concise form. Each title features an introduction
placing the book in context and highlighting
its special contribution to the advancement of
human understanding.

See our expanding range at
flametreepublishing.com